THE ENCYCLOPEDIA OF

CLASSIC CARS

SUPER-CHARGED

THE ENCYCLOPEDIA OF CLASSIC CARS

Chris Harvey

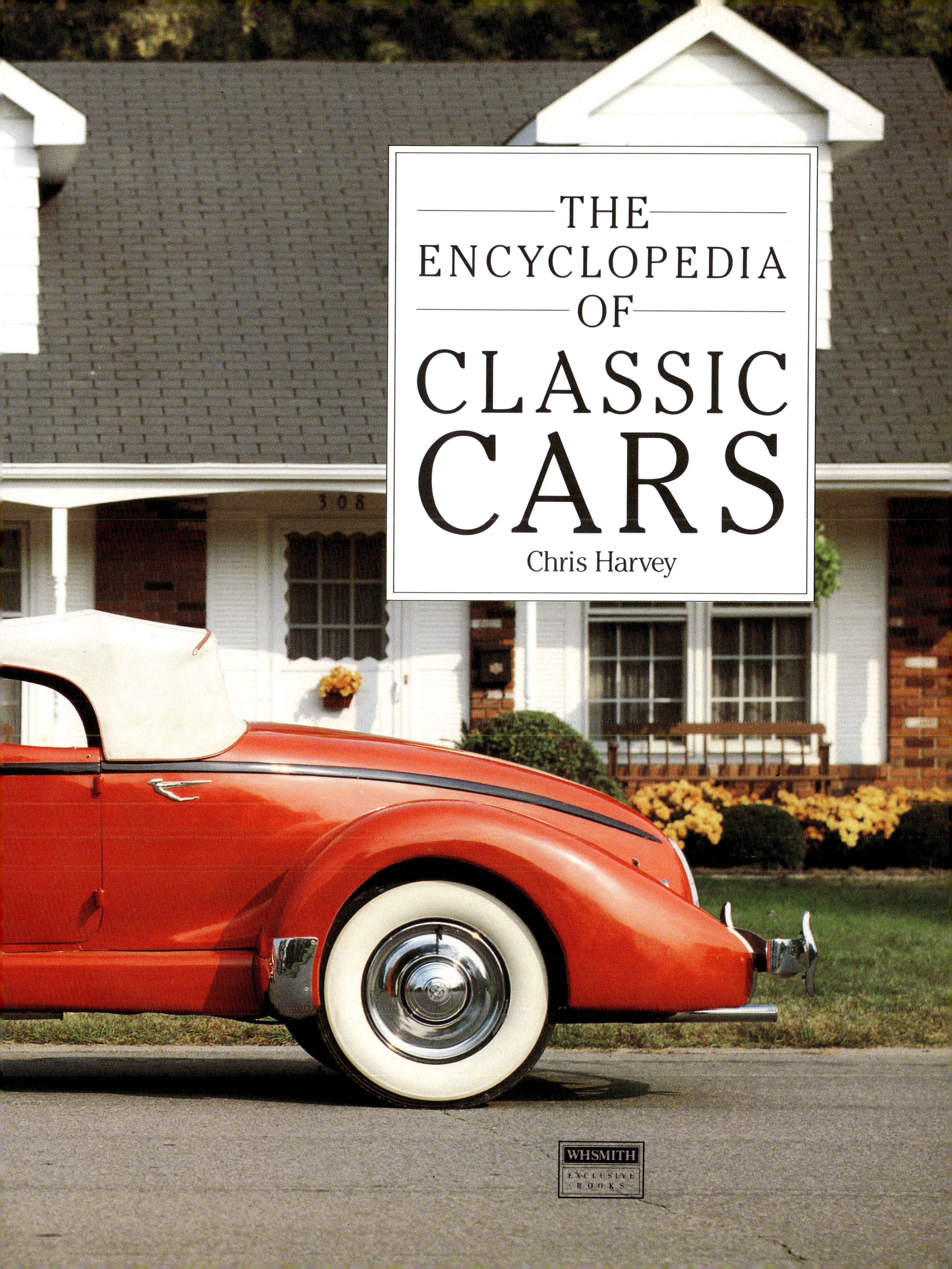

WHSMITH
EXCLUSIVE BOOKS

This edition published in 1987
exclusively for W H Smith
by Octopus Books Limited
59 Grosvenor Street
London W1X 9DA

ISBN 0 7064 2874 9

Produced by Mandarin Publishers Limited
22a Westlands Road
Quarry Bay, Hong Kong

Printed in Hong Kong

Page 1: Alfa Romeo's 1750GS Zagato

Pages 2-3: Auburn's Boat-Tailed Speedster.
Owner: Auburn-Cord-Duesenberg Museum.

These pages: Jaguar's SS100.

CONTENTS

INTRODUCTION

It is no simple matter to define a classic car. With more than a million models produced by 5,000 manufacturers over the past 100 years, it is inevitable that some candidates will fall by the wayside and only the greatest will survive. Nor can age alone be a qualification. Only when a particular model has proved itself to be far in advance of its contemporaries and of anything which went before, can it be considered a great classic.

Many classic cars are within the grasp of the ordinary motoring enthusiast, who might be surprised to find what a modest purse can buy. A modest budget, however, effectively rules out many pure competition cars, as these have become part of a booming antique industry. In any case, one of the great joys of owning a classic car is being able to drive it on the road, and this disqualifies many racing cars.

Perception of what makes a car a classic also tends to be influenced by locality. It is easy for Europeans to overlook the great American marques and for Americans to be blinded by the glittering chrome of domestic products – and hardly anyone gives Japanese models a chance.

Collecting classics can attain the status of a science, dependent on depth of pocket, space for storage and an intimate knowledge of exactly what constitutes a classic. For the impecunious enthusiast who dreams of owning a vintage car, the Austin Seven can represent the greatest practical investment, in terms of both money and pleasure. Such machines cost little to buy, even less to run and occupy only a tiny parking space, they can be extremely rewarding to own and drive, and are powerfully evocative of a bygone era.

For the more ambitious collector, classics are easier to identify. The sporting scene is no better served than by the magnificent Blower Bentley or by the far more lithe, but less awesome, Alfa Romeo 6C 1750GS Zagato. Prices are high, however, partly because of problems of supply: it may be necessary to wait several years before acquiring such works of art. Bugattis of all shapes and sizes are attaining the status – and price – of old masters and the post-war equivalent, Ferrari, is heading the same way.

In general, sports cars are far better investments than saloons, simply because they are more glamorous, generally occupy less space and certainly cost far less to maintain and restore than machines featuring vast, opulent bodywork. With the notable exception of Duesenberg and Stutz which rival such great marques as Mercedes, the pre-war American classics, such as the 16-cylinder Cadillac 452 and Packard Twin Six, are chiefly touring cars of immense stature. While the British Rolls-Royce once laid claim to the title of 'The Best Car in the World', these American giants are now considered its equal in the classic stakes.

Post-war classics can be more difficult to spot but in fact are easier to collect, because they are usually more plentiful and cheaper as a result. For the budget-conscious collector, the British Austin marque is again a front runner, such is the inspired styling of the Healey models, large and small. Restoration costs can be high, because post-war cars tend to rely more heavily on mild steel in their construction, and this rusts away in climates damper than, say, southern California. The post-war sporting market is dominated by Jaguars, however, ranging from the cheaper XK models to the Le Mans-winning D-types, some of the last racing cars that could be driven on the road. Some machines are the equals of all but the most expensive Ferraris in value and are generally much more practical to use because the mechanical components – notably the twin-overhead camshaft XK engine – were shared with the baseline models. The rarest of the rare, such as the Spanish Pegaso and the supercars of the 1960s and 1970s, will remain the ultimate investments. The Lamborghini Miura and Countach are worth looking out for, and Porsche occupies the mid-range with the 911 Carrera RS, slightly ahead of Jaguar's cheaper E-type.

Locating classics need not be a problem, providing you know what you want: each marque, and frequently each model, has its own following, crystallized in clubs and these can be tracked down in magazines devoted to the subject. Membership opens the door to an unrivalled source of information on what is available, and how to preserve and restore your investment, as well as providing enjoyment in the form of competitions. These may vary from race meetings to *concours d'élégance* which establish the best example of the marque or model.

Originality is all the rage – and has been for years – when it comes to preservation and restoration. The aim is generally to return the car in question to as near as possible the original specification and appearance. It is now considered a crime to 'over-restore' – chrome-plating parts which would normally have been painted, for instance – and a new trend is developing which corresponds with a move in the world of antique furniture. The most intelligent restorations now try to retain a car's patina, reasoning that if a classic is old, it should *look* old and elegant, and the owner should not try to reproduce a brand-new look. As much of the original material as possible is retained and, if a car shows the scars of a long and active life, it is repaired gently. New pieces of metal are let in only where vital rather than a complete new panel being made, even though that might be cheaper. Where less than 20 years ago original rivet heads would have been considered unsightly and would have been ground away and covered in by welding, they are now retained as a special feature.

Spotting future classics is harder than finding existing ones. In essence, the car under surveillance has to have extraordinary character, plus a specification which puts its performance – not necessarily outright speed, but quite possibly handling, too – and appearance far ahead of its rivals. Values are also affected by availability, the absolute classics coming from limited runs which will never be repeated.

Taking classic cars era by era, the pre-First World War Hispano-Suiza Alfonso is only now emerging as one of the greatest, not just because it was the world's first sports car, but also because it is pleasant to drive, whereas many of its contemporaries certainly were not. The Mercedes Benz SSK of the 1920s is conquering new fields now that historic racing is swinging back towards road events, (such as the Mille Miglia in Italy), and the spine-chilling scream of its mechanically engaged supercharger can be guaranteed to thrill the crowds.

△ ***Ferrari's 365 GTB/4, popularly known as the Daytona.***

Almost every Bugatti is great, but none can beat the Type 57SC Atlantic coupé, of which only three were made. Enthusiasts always knew it was great, but values are soaring even higher now that the whole world has been let in on the secret. Cars of the 1940s are scarce because of the Second World War, and in this context the Jaguar XK120 is set to reach great heights not only as one of the most beautiful sports cars ever made, but also as one of the most practical, even though it has the world's first grand prix-style, twin-overhead-camshaft, hemi-head engine.

Jaguar's D-type sports racer will surely prove to be the greatest classic of the 1950s above all else because nothing will ever match its distinctive aeroplane-inspired shape. Competition is severe in the 1960s but Ferrari's last great front-engined road racer, the 250GTO, is already established well in front of the classic field, with the Ferrari-inspired AC Cobra set to leap forward again alongside the first mid-engined supercar, the Ford GT40 and the V12-engined car it inspired, the Lamborghini Miura. American classics are likely to be led by the awesome power outputs of the 1966 Chevrolet Corvette Stingray and the Dodge Charger.

The Porsche Carrera RS and Lamborghini's Countach dominate the 1970s, alongside Ferrari's last great front-engined grand tourer, the Daytona. There is a wider range of cars from which to choose in this decade, because cars are younger and in better condition, and so there are still bargains to be found: Porsche's 911S, produced immediately before the Carrera RS, differs only in detail and has a purer shape, yet it has only just reached half the price of its successor.

The early years of the 1980s belonged to the Audi Quattro, such was the quantum leap it offered in performance and in its strange, brooding, colour-keyed styling. As the decade progressed, the most inspiring cars turned out to be those outlawed from international rallying because of their enormous, grand prix-style power outputs and four-wheel-drive adhesion. Of these, the MG Metro 6R4, sold off at very low prices by a financially beleaguered manufacturer, is the potential classic, not only because of its amazing performance, but because it has an exclusive appeal – the exposed mechanical features started a styling trend and made the MG 'The World's Ugliest Supercar'!

ABARTH

Ordinary cars festooned with go-faster goodies are seldom regarded as classics . . . unless they bear the legendary name Abarth. Ordinary cars invariably remain quite ordinary under the tinsel, but the products of Karl Abarth were extraordinary because he really understood the tuning business which spawns such creations.

Abarth's cars invariably used mass-produced components – such as an engine or a floorpan – from popular cars because it would have cost too much to make the parts from scratch, but they were frequently modified with such ingenuity that the end product was a very individual machine.

△ ***Cars as fast as the Fiat-Abarth 1000 Turismo Competizione Radiale are monuments to tuner Karl Abarth.***

1000TCR

YEARS MADE: 1967–71.
ENGINE: In-line four-cylinder, 982 cc, twin overhead camshafts, 105 bhp at 8000 rpm.
FRAME: Integral, based on Fiat 850.
PERFORMANCE: 118 mph (190 km/h).
CURRENT PRICE GUIDELINE: * * Very rare, good examples difficult to locate, except in Californian-style climates.
PROBLEMS TO LOOK FOR: Check that components are original and numerous special parts are in good order. Many bogus examples built from hotch-potch of Fiat and Abarth parts.
CLASSIC CAR STATUS: * * * Likely to improve in value as cars become scarcer.

Abarth's pedigree was immaculate: before the 1939–45 war he was a champion motor-cycle racer, and afterwards he was instrumental in getting fellow Austrian, Ferdinand Porsche, ransomed from jail in France, where he had been confined for doing war work on German tanks. The ransom money was raised by the Porsche family who designed a Cisitalia Grand Prix car for Italian industrialist Piero Dusio (see page 49). When cash for this over-ambitious project ran out, Abarth reorganized the Turin-based workforce to produce much-needed spare parts, such as exhaust systems for Fiats.

Funded by silencer profits, he built his first Fiat-based competition car in 1949, and never looked back, selling high-performance accessories to the owners of ordinary Fiats: everything from scorpion badges (Abarth was born under the sign of Scorpio) to cast-aluminium sump pans that hung out of the back of these rear-engined cars, making them look very powerful. Even if such tiny Fiats had only ordinary engines, they sounded special thanks to their Abarth exhaust systems, the driver's brio making up for the rest.

During the 1960s, Abarth enlarged standard Fiat engines to an amazing degree – to as much as 1000 cc from 600. It is hard enough to add an extra 400 cc to an engine of nearly 4 litres, but no mean feat if you have only four cylinders to play with, totalling little more than half a litre. Suffice to say, no tricks of the tuning trade were unknown to Abarth and he produced everything from long-stroke crankshafts that needed the block machining out to a perilously thin degree, to twin-cam cylinder heads with tortuous timing chains. But although he produced one-off competition cars for publicity, almost always he chose small, popular cars for his main production, because they were cheaper to start with and he could therefore spend more on his performance equipment.

The first real production Abarth was the front-engined 207A Spyder of 1955, based on a Fiat Millecento. He followed this up by concentrating on rear-engined Fiat 600-based cars (notably the Zagato-bodied 750 with 'double bubble' roof to lower the overall height) until 1961, although there were a few coupés using Alfa Romeo Giulietta components in 1958, and a special run of lightweight bodies for Porsche GT cars in 1960.

Abarth also developed Simca 1000-based cars in 1961 alongside the normal run, until closer links were forged with Fiat in 1966. After that, some of the most memorable Abarths appeared, such as the 1000TCR (for Turismo Competizione Radiale). These formidable machines, which started life as 28 bhp Fiat saloons, and produced as much as 110 bhp for the European Touring Car Championship, came complete with rear wing, bulging wheel-arches and huge front oil and water radiators to help balance a much-modified and highly sophisticated hemi-head engine.

Such outrageously fast cars as the 1000TCR remained as a monument to Karl Abarth after he sold out to Fiat in 1971 and retired.

Hardly any two Abarths were exactly the same. Partly because of this, partly because of their amazing performance, and partly because they are still relatively cheap, they have become highly collectable. The main problem when buying an Abarth is making sure that it is a genuine car, because far more ordinary Fiats (and to a certain extent Simcas) received Abarth conversions than were bought in by the factory for Abarth production. It is always vital, therefore, to check that a potential purchase has an Abarth chassis plate that tallies with numbers stamped in the body panels and that the number of its engine is also on the chassis plate. In common with many ageing Italian products, Abarths suffer seriously from corrosion, and many parts special to Abarth may now be almost unobtainable; many that were common to Fiat, however, are easily bought!

SUMMARY

Plus points

Rarity value.

Tremendous performance, faster than its more common rival, the Mini-Cooper S.

Appreciates in value providing special parts are not damaged.

Minus points

Prone to extensive corrosion in damp climates.

Fragility of mechanical parts.

Many bogus examples can trap the unwary.

Needs racing-car-style maintenance.

AC

The AC Cobra has rapidly become acclaimed as one of the most emotive post-war sports cars because of its classical shape and stunning acceleration. It was an immediate success when it appeared in 1962 and has since been much copied in the so-called replica market, although only Autokraft's Mark IV, produced on the original jigs, has been allowed to be named Cobra.

COBRA

YEARS MADE (Mark II): 1963–5.
ENGINE: V8, 4735 cc, overhead valves, pushrods, variable tune, average 300 bhp at 5750 rpm.
FRAME: Twin tube.
PERFORMANCE: 153 mph (246 km/h).
CURRENT PRICE GUIDELINE: ****
Approximately 1000 made 1962-8, in great demand. Mark IV models in ready supply.
PROBLEMS TO LOOK FOR:
Mistreatment, aluminium body easy to damage, expensive to repair, accident-damaged chassis tubes.
CLASSIC CAR STATUS: ***** One of the most emotive post-war sports cars.

The AC Cobra's body was developed by a quite circuitous route from the historic barchetta ('little boat') open two-seater created by the Italian firm Touring on Ferrari sports racing chassis from 1948. It was during this period that the British firm Cooper began building successful tubular chassis for competition, which inspired many similar creations. These included one by the little-known British engineer, John Tojeiro, for racing driver Cliff Davis. At the same time, Davis commissioned a body for his Tojeiro that followed the lines of the barchettas as closely as possible – and then enjoyed a lot of success on the race track with his new car.

An old-established engineering firm at Thames Ditton, Surrey, which built AC cars in small quantities as a sideline, was searching for a new design at the time and reached an agreement with Tojeiro to produce his car on a royalty basis. This was the AC Ace which appeared in 1954 with AC's own six-cylinder engine in place of the rather tall Bristol unit which had powered the Tojeiro. However, the Bristol –

▽ The classic shape of the original AC Cobra lives on in the Mark IV version constructed by Autokraft.

although far more expensive than the venerable AC engine – was so much more powerful that it became a popular option from 1956. Its future looked doubtful, however, when Bristol stopped using it for its saloon cars in 1961 and opted for a cheaper and more powerful American V8.

The virtues of such American engines had not been overlooked on that side of the Atlantic, either. Enthusiasts in the United States – which represented by far the largest market for European sports cars – had for years been substituting home-grown units for the original engines as a cheap source of new power. It needed only an entrepreneur, Texan former racing driver Carroll Shelby, to order an AC Ace body and chassis unit modified to accept a Ford Fairlane engine, for the Cobra to be created, in 1962.

The combination of the AC chassis and Shelby's 4.2-litre engine was a particularly happy one because Tojeiro had originally designed the chassis with all-independent suspension and the engine set well back (as the Bristol unit was so tall), which gave the car exceptionally forgiving handling, even with the heavier V8 engine in place. AC had already uprated the lightweight chassis with disc brakes, which coped well with the more or less doubled power, and the AC Cobra was an immediate success.

The first cars were sold by Shelby, who installed the power trains. In 1963 the capacity was increased to 4.7 litres for even more performance. AC, meanwhile, improved the chassis by replacing their worm and sector steering with the more precise rack and pinion originally used by Tojeiro, and strengthening the suspension for a Mark II model in the same year. In this form, the Cobra 289 (named after the engine's US-style cubic-inch displacement) became a great success, not only on the road, but also on the track, with tuned versions producing as much as 330 bhp to make it one of the fastest sports cars of its day.

Ford also played a large part in marketing the Cobra because, although the handbuilt car could be produced only in small numbers, it helped to give the company a youthful, new, high-performance image. Racing versions took the world championship for GT cars from Ferrari in 1965, and soon Ford and Shelby were developing an even more powerful version, the Mark III 427, with a 7-litre V8, strengthened chassis, and more modern full wishbone and coil suspension. Much wider wheels were also fitted, which – with bulging wings – made the car look even more brutal. And it lived up to its appearance, too: performance was shattering, with a 0-60 mph (0-97 km/h) time below 5 seconds, and for years it was the fastest-accelerating sports car in the world.

Ford moved on to concentrate on its own mid-engined GT40 from 1966, and AC had to give up producing the Cobra in 1968 when new American environmental laws outlawed this, and several other more traditional sports cars, from their main market. But numerous reproductions appeared on the home market, including some excellent examples made in Weybridge by a firm called Autokraft. Eventually this company managed to take over the original AC jigs, to produce a modernized 5-litre version called the Mark IV, from 1984.

New cars cost about £40,000 in 1986, with original examples fetching more because of their antique status. Of the early Cobras, the 289 is the best to drive because it is far more nimble than the muscle-bound 427, although the bigger-engined car has its following purely because it is the most powerful. With the very similar Mark IV still in production, there is no problem with spare parts for what is essentially a very simple and durable car. But many Cobras were hacked about during the 1970s when they were not so much appreciated and it is wise to check originality – so vital to a classic car's value – when you are buying an old model.

SUMMARY

Plus points

Appearance.

Performance (although 7-litre is rather unmanageable).

Availability of spare parts.

Minus points

Strong competition to buy original examples.

Delicate body panels.

ALFA ROMEO

Badges bearing the name Alfa Romeo Milano have long been associated with some of the most exciting cars on the road. Much of the credit must be given to the craftsmen who produced them, who are recognized as some of the most skilled metalworkers in the world. It is significant that whenever Alfas have been produced by less-skilled workers outside the confines of the manufacturing base in Lombardy, northern Italy, the firm has been in trouble.

This was certainly the case at the turn of the century when the fledgling company started out by assembling French Darracq cars in Naples. The product was of such poor quality that the firm had to move north in 1909 to be re-organized as the Anonima Lombarda Fabbrica Automobili (hence Alfa) at Portello, on the outskirts of Milan. It was subsequently taken over by industrialist Nicola Romeo in 1915, at which point the name Alfa Romeo was born. Its chief designer, Giuseppe Merosi, was talented enough to produce the RLTF model which won the Targa Florio in 1923, but it was not until chief test driver Enzo Ferrari persuaded Fiat's Vittorio Jano to replace Merosi at Portello that the legend was established.

1750GS ZAGATO

YEARS MADE: 1929–33.
ENGINE: In-line six-cylinder, 1752 cc, twin overhead camshafts, 85 bhp at 4500 rpm.
FRAME: Twin girders in ladder formation.
PERFORMANCE: 90 mph (145 km/h).
CURRENT PRICE GUIDELINE: ***** Very rare, good examples well documented. May be necessary to wait years for right one to come on market.
PROBLEMS TO LOOK FOR: Bogus claims to competition history, ageing components.
CLASSIC CAR STATUS: ***** One of the greatest ever and a gilt-edged investment.

The first Alfa Romeo created by Jano – subsequently hailed as the greatest Italian car designer – was the sensationally successful P2 Grand Prix car,

the forerunner of the 1750GS. The P2 combined a startling performance with exceptionally good roadholding. From that point, handling was always to be Alfa Romeo's forte.

The P2 formed the basis for Jano's first touring and sports Alfas featuring highly advanced power units, as befitted a factory which was now also renowned for its aero-engines. The six-cylinder units used in cars were made first in 1500 cc, and then in 1750 cc forms and it was the twin-cam supercharged 1750GS with ultra-lightweight bodywork by firms like Zagato that captured everybody's imagination while winning most of the world's top road races.

These cars were followed by the even more exotic straight eight-cylinder 8C 2300 series that dominated the Le Mans 24-hour race and the Mille Miglia

△ The 1750GS Alfa Romeo with lightweight Zagato bodywork fired imaginations everywhere. Owner: Alfa Romeo Museum, Arese.

round-Italy road race during the early 1930s. Sadly, such machines cost too much to manufacture during a time of economic depression, and Alfa Romeo had to be rescued by the Italian government to survive. But little changed, and the state-owned Alfa Romeo continued to produce fast, prestigious sports cars, together with Grand Prix models to compete against the might of German Mercedes and Auto Unions, including the fearsome twin-engined Bimotore of 1935, developed by Scuderia Ferrari (the Alfa works team run by Enzo Ferrari), with one P3 engine in front of the driver and another behind him. Both engines drove a centrally-mounted three-speed gearbox which then powered the rear wheels. When it ran properly, the Bimotore was capable of 200 mph (320 km/h), but the team could never cure its enormous appetite for tyres, which wasted the car's extra performance. Such monsters emphasized that the smaller Alfas, like the 1750GS Zagato, really were the greatest.

SUMMARY

Plus points

One of the most beautiful between-the-wars sports cars.

A delight to drive.

Durable if well maintained and treated with reasonable sympathy.

Minus points

Great cost of having spare parts made.

△ ***The Alfa Romeo Giulietta Sprint – one of Satta's greatest designs.***

GIULIETTA SPRINT

YEARS MADE: 1954–62.
ENGINE: In-line four-cylinder, 1290 cc, twin overhead camshafts, 103 bhp at 6500 rpm.
FRAME: Integral.
PERFORMANCE: 112 mph (183 km/h).
CURRENT PRICE GUIDELINE: * Relatively rare as many examples have been scrapped. Good examples difficult to locate except in kind climates.
PROBLEMS TO LOOK FOR: Extensive corrosion, worn-out mechanical and electrical components in hot climates, unobtainable trim.
CLASSIC CAR STATUS: * Likely to improve in value as cars become scarcer.

Most of the original factory at Portello was destroyed during the 1939–45 war and Alfa had to resume production with cars based on pre-war models. These were quite specialized cars that could be built in relatively small quantities.

But soon after, everything changed. A new designer, Orazio Satta, produced the brilliant little Giulietta in 1954, and demand reached such heights that Alfa Romeo became established as a volume car producer. The Giulietta's 1300 cc five-bearing, four-cylinder engine became one of the world's very first mass-produced, twin-overhead-camshaft units, endowing this beautifully balanced car with a superb performance even in saloon (or Berlina) form. In true Alfa Romeo tradition, the Giulietta appeared in all manner of guises, one of the most notable being the Bertone-bodied two-door Sprint.

SUMMARY

Plus points

Brilliant design.

High performance.

Nimble handling.

Still relatively cheap.

Appreciates in value if body and chassis can be preserved.

Minus points

Prone to terminal corrosion in damp climates.

Tendency to be mistreated by the uninitiated.

Twin-cam engine and transmission need careful maintenance and handling.

GIULIA SPRINT SPECIALE

YEARS MADE: 1963–5.
ENGINE: In-line four-cylinder, 1570 cc, twin overhead camshafts, 112 bhp at 6500 rpm.
FRAME: Integral.
PERFORMANCE: 130 mph (209 km/h).
CURRENT PRICE GUIDELINE: ** Rare. Good examples difficult to locate, except in Californian-style climates.
PROBLEMS TO LOOK FOR: Extensive corrosion, worn-out mechanical and electrical components in climates where body and chassis have survived, unobtainable trim.
CLASSIC CAR STATUS: ** Likely to improve in value as cars become scarcer.

Following the spectacular success of the Giulietta, Alfa developed further notable variations on its basic theme, including the Giulia and the Pininfarina-bodied, open two-seaters, called the Spyder, and the Spyder Veloce. The latter was introduced with an uprated engine in 1957. Further juggling of engines, chassis and bodies resulted in Bertone's 114 bhp Sprint Speciale in 1959, followed by the SZ (for Sprint Zagato) in 1960 with an ultra-lightweight Zagato fixed-head competition body.

SUMMARY

Plus points

Stunning appearance and performance.

Very good handling.

Economy of operation once in good condition.

Possibility of rapid rise in value.

Minus points

Prone to terminal corrosion in damp climates.

Twin-cam engine and transmission need careful maintenance and handling.

High risk of damage to extremities.

DUETTO SPIDER

YEARS MADE: 1966–7.
ENGINE: In-line four-cylinder, 1570 cc, twin overhead camshafts, 109 bhp at 6000 rpm.
FRAME: Integral.
PERFORMANCE: 115 mph (185 km/h).
CURRENT PRICE GUIDELINE: ** Early models are rare, later ones also desirable. Beware of examples in poor condition, better to wait for a good one as restoration costs are very high compared to ultimate value.
PROBLEMS TO LOOK FOR: Extensive corrosion, worn-out mechanical and electrical components in climates where body and chassis have survived, some trim difficult to obtain on early models.
CLASSIC CAR STATUS: ** Likely to improve when new cars become obsolete.

Eventually, the basic power unit was enlarged to 1570 cc for the new series of Giulia models from 1962, with the Sprint Speciale and a redesigned Spider, the Duetto, as the most spectacular.

Although the Duetto lasted only a couple of years before it was uprated to 1779 cc, the same type of bodyshell continued to be used on open Alfa Romeos for two decades.

Versions sold in America inevitably ran into problems with exhaust emission regulations, with the result that the glorious European Weber carburettors had to be replaced with fuel injection.

The body – although it retained the same basic shape – came in for a good deal of revision in 1970, especially at the back, where it received a lopped-off tail. This conformed to a theory prop-

△ The spectacular Sprint Speciale by Bertone was one of the most attractive Alfa Romeo coupés. Owner: Lincoln Small.

◁ The original Duetto donated its lines to one of the longest-lasting series of Alfa Romeo spyders. Owner: Barry Coupe.

agated by Kamm, who discovered that if the airstream around the top deck of a car left the tail virtually horizontally, the tail could be squarely chopped off, reducing the overhanging weight without spoiling aerodynamics. (This theory was demonstrated enthusiastically by the contemporary TVR sports cars – see page 156 – as well as racing machines such as the Ferrari 250GTO and Ford GT40 – see pages 68 and 75.)

Further revisions to the fuel injection system extracted more power and led to the American Spider being called the Veloce (or 'speed'). But no matter what it was called, Duetto, Spider or Veloce, the small, open Alfa remained a long-time favourite.

SUMMARY

Plus points

- Attractive appearance.
- Good performance and handling.
- Open two-seaters becoming rare.

Minus points

- Prone to terminal corrosion in damp climates.
- Twin-cam engine and transmission need careful maintenance and handling.
- High risk of damage to extremities.

MONTREAL

YEARS MADE: 1970–5.
ENGINE: V8, 2594 cc, four overhead camshafts, 230 bhp at 6500 rpm.
FRAME: Integral.
PERFORMANCE: 137 mph (220 km/h).
CURRENT PRICE GUIDELINE: * * Really good examples difficult to locate, average examples more common.
PROBLEMS TO LOOK FOR: Neglect, especially of power train, corrosion in damp climates.
CLASSIC CAR STATUS: * * Likely to improve.

Alfa Romeo's organization, Autodelta, was set up to handle a competition programme for lightweight examples of the Giulia so that factory personnel would be free to concentrate on production models. Autodelta, under former Ferrari engineer Carlo Chiti, enjoyed so much success with production-based machines that it was decided to go ahead with a full-scale sports racing car for 1967: the Tipo 33.

This followed what had become conventional mid-engined lines, but with a Chiti-designed four-overhead-cam hemi-head V8 engine and six-speed gearbox. In general, these cars struggled against better-financed opposition from Ferrari, Porsche and Ford, but a roadgoing version of the 1967 V8 engine was used in a new Bertone coupé in 1970. This was called the Montreal after a prototype shown at the Canadian World Exhibition in 1967. Although the exotic two-seater cost twice as much as any other Alfa Romeo, it sold comparatively well for the next five years, re-establishing the Milanese firm's prewar reputation for making supercars.

SUMMARY

Plus points

- Rarer than a Porsche 911.
- Similar performance to cheaper Porsches, more forgiving handling.

Minus points

- Exotic engine needs exacting maintenance.
- Spare parts scarce.
- Prone to corrosion.

▽ The Montreal became the last Alfa Romeo supercar: its V8 engine also powered sports racing cars. Owner: The Patrick Collection.

ALFASUD

YEARS MADE: 1972–82.
ENGINE: Flat four-cylinder, early models 1186 cc, later up to 1490 cc, single overhead camshaft, early models 73 bhp at 6000 rpm, later up to 105 bhp at 6000 rpm.
FRAME: Integral.
PERFORMANCE: Early models 93 mph (150 km/h), later models 115 mph (185 km/h).
CURRENT PRICE GUIDELINE: * Cars in good condition can be a bargain because majority typically suffer from early terminal corrosion, which can be halted with care.
PROBLEMS TO LOOK FOR: Extensive body corrosion, general neglect.
CLASSIC CAR STATUS: * Sprint models will improve after, perhaps, ten years.

Once it became a state-owned operation, Alfa Romeo was subject, naturally, to considerable political influence and became involved in government plans to set up a new factory in southern Italy to supplement the operation in Milan. It all seemed so logical: Alfa Romeo would return to Naples to produce a new small car in an area where unemployment was high, and wages could be held at a lower level than in the traditional car-building areas of the North. In this way, the Italian government would save money both on supporting the unemployed and producing a profitable new car – let alone benefiting a community deprived of sufficient job opportunities. The result was the Alfasud ('Alfa South') launched from a new factory with a new workforce in 1972.

As ever with Alfa Romeo, the design was advanced and to an excellent concept. It was the work of a team led by an Austrian, Rudolf Hruska, who had arrived in Italy to work on the Cisitalia racing car project with Karl Abarth (see page 8) after having been involved in the original Volkswagen project with their countryman, Professor Ferdinand Porsche.

The Alfasud's single-overhead-cam, flat four-cylinder engine offered a typically high standard of Alfa Romeo performance in a lightweight bodyshell for what was to be the firm's first front-wheel-drive car. The suspension, too, was different from existing models: Ford-style MacPherson struts were used at the front, and a rigid axle was carried

△ ***The Alfasud represented a brilliant small car concept in spite of massive production problems. Owner: Vernon Thompson.***

on coil springs and located by trailing arms at the back.

The theory was brilliant, but the execution – partly due to the problems in turning an unskilled local population into car workers – was not. Sales were good at first until it was discovered that early examples of the Alfasud were so badly made that they lasted only a relatively short time. Within a few years Alfa Romeo had lost the best part of a reputation that had taken ten times as long to establish. It sowed the seeds for a rapid decline in the company's fortunes although, ironically, improved versions of the Alfasud sold well and the model continued in production in one guise or another into the 1980s.

Eventually, declining sales of its volume products led to Alfa Romeo being taken over by Fiat to produce quality cars once again.

SUMMARY

Plus points

Superb concept.

Good performance and handling, faster than a Mini.

A classic if you can store it rust-free for long enough.

Minus points

Appalling build quality of many examples.

Terminal corrosion.

ALLARD

The cars built by Sydney Allard were really the forerunners of the AC Cobra (see page 9). As such they were among the first to demonstrate the advantages of using a cheap and powerful American V8 engine in a lightweight European chassis, which could provide far better handling than the cars built in the United States at the time.

J2X

YEARS MADE: 1951–3.
ENGINE: (typically) Cadillac V8, 5394 cc, overhead valves, pushrods, 160 bhp at 3800 rpm.
FRAME: Tubular.
PERFORMANCE: 111 mph (188 km/h).
CURRENT PRICE GUIDELINE: *** Very rare, anticipate a long wait to find one for sale at a reasonable price.
PROBLEMS TO LOOK FOR: Authenticity. Check that all components, including the chassis, are original. The earlier J2 is not so valuable, and later power units are likely to have been fitted.
CLASSIC CAR STATUS: **** Certain to maintain a good value.

Sydney Allard, who ran a Ford garage in south London, was one of the leading drivers in reliability trials (the equivalent of today's rallies) before the Second World War and quite naturally used machines based on Ford components. The half-dozen cars built before 1939 usually took the form of a short-wheelbase (and, therefore, lighter) Ford V8 fitted with a skimpy sports body – in the first instance, from a Grand Prix Bugatti!

One car was made with a V12 engine, however, and ran well in a production sports car race at Brooklands. The way in which it was qualified for such an event was typical of the delightfully informal spirit of the times. Although it was patently not in normal production, it was allowed to compete because it was a prototype and as such could hardly be expected to last the race!

Normal Allards were so successful during the heyday of trials in the three years before the Second World War that there was still a considerable demand for similar machines for road and occasional competition use when hostilities ceased. Britain was also starved of new cars in the immediate post-war era, so Allard decided to put his machines into small-scale production.

Initially, Allards were available in spartan, short-wheelbase, two-seater, J-type competition form; as a medium-wheelbase, touring K-type; or as a long-wheelbase, four-seater, open L-type; as a more luxurious, drophead coupé M-type; or as a two-door, four-seater saloon P-type. Allard had his own pressed-steel chassis made and fitted them with either Ford or Mercury V8 engines and Ford-based suspension using transverse leaf springs. The front axle was sawn in half and fitted with pivots to provide a primitive form of independent front suspension. The bodies were handmade with a distinctive 'waterfall' radiator grille.

These relatively light cars rapidly gained a reputation for high performance, as Allard developed a more sophisticated J2 competition version with coil spring suspension, including a de Dion rear axle, for 1950. Import restrictions hindered the fitting of the best American engines, so many chassis and body units were shipped to the United States where formidable power units like Cadillac's 160 bhp V8 were fitted, making the J2 an immediate race-winner. One such car, driven by Sydney Allard and American Tom Cole, finished third in the 1950 Le Mans 24-hour race.

The J2X was then developed for 1951 with a more modern tubular chassis, but retaining the same basic layout as its brutal predecessor – including the by-now old-fashioned cycle-wing body. In this form, with Cadillac's engine producing a tremendous amount of torque at low revs, the J2X was ideally suited to the tight airfield circuits popular in American club racing at the time. But once the car was travelling at more than 100 mph (161 km/h), the primitive aerodynamics inhibited any further acceleration, no matter how much power was extracted from the Cadillac engine or from the more advanced Chrysler 'hemi-head' V8s fitted later. International regulations were also about to outlaw such vestigial bodywork on sports cars in favour of 'proper' full-width bodies, rendering the J2X ineligible for the top competitions.

The same strictures did not apply to the saloon, which had fairy conventional bodywork, and in 1952 Allard won the Monte Carlo Rally with a P-type.

Despite such success, however, sales soon faltered because the rival Jaguars, which were far more readily available, were not only cheaper and faster but also looked much more attractive. Allard attempted to counter the slump by revising the J2X as the J2R

sports racing car in 1953 with modern full-width body, but the demand had fallen so much in the face of Jaguar's C-type (see page 86) that only seven were built, and within a few years Allard virtually gave up making cars to concentrate again on his Ford dealership and selling performance equipment.

The J2X remains as one of the last survivors of an era which produced really ferocious sports cars that paid little heed to aerodynamics, or to any notion of creature comfort.

SUMMARY

Plus points

Rarity value.

Startling acceleration.

Brutal appeal of a bygone era.

Very durable.

Minus points

Expense of restoring aluminium body.

Difficulty in finding original mechanical components.

△ ***Sydney Allard's sports cars – of which the J2X was the prime example – were the last survivors of an era in which the comfort of the occupants was of less concern than the outright performance.***

ALPINE

For the best part of 30 years, Renault-powered Alpines have been France's only sports cars, and until recently the A110 was the best of them all. The Renault-Alpine GTA V6 Turbo, launched in 1986, continues the firm's traditions and has looked set from the start to become another of the motoring world's most memorable classics.

A110

YEARS MADE: 1969–71.
ENGINE: In-line four-cylinder, 1565 cc, overhead valves, pushrods, 138 bhp at 6000 rpm.
FRAME: Tubular backbone with integral fabricated subframes front and rear.
PERFORMANCE: 134 mph (215 km/h).
CURRENT PRICE GUIDELINE: * * Very rare outside Europe.
PROBLEMS TO LOOK FOR: Mistreatment, check that engine and transmission are original specification for model.
CLASSIC CAR STATUS: * * Likely to improve in value.

The early cars produced by Jean Rédelé, a Renault dealer and former racing driver from Dieppe, were insignificant chiefly because they were powered by quite mundane engines – first from a 4CV, later a Dauphine – and were based, initially, on a standard saloon-car floorpan. The Alpine really started to come of age when, in 1957, Rédelé made his own chassis for the A108. It was a simple affair with a large steel tube for a backbone, clothed with striking new bodywork designed by Marcel Hubert. An open version was listed, but virtually every Alpine became a small, bug-like coupé.

The A108 turned into the A110 and became a classic in 1963 when it was fitted with the Renault 8 engine, which had far more tuning potential. During the next 15 years, the capacity was to go up from 998 cc to 1800, and the power from 45 bhp to 185, with all manner of cylinder heads, some of the most notable being hemispherical designs by Gordini. By the time the A110 – or Berlinette as it was known – was up to 1300 cc in 1967, it was fast enough to challenge Porsche for victory in the Monte Carlo Rally. Like the winning Porsche 911, the Alpine had superb traction because its engine was mounted behind the rear axle line.

Its performance was increasingly impressive despite the relatively small engines because the flimsy glass-fibre bodywork was extremely light, giving the A110 – with its very rigid chassis – an overall weight of around 1400 lb. And when the A110 received a 1600 cc, 138 bhp version of the Renault R16 engine it became a world-beater, taking the first three places in the Monte Carlo Rally in 1971 and 1973, besides winning numerous other top events.

SUMMARY

Plus points
- High performance.
- Entertaining to drive.
- Durable and relatively easy to repair.
- Good supply of spare parts at present.

Minus points
- Can be difficult to drive for the uninitiated.
- Very noisy and cramped.

▽ ***The tiny rear-engined Alpine A110 – particularly this 1969 example – was as nimble as a mountain goat.***

RENAULT-ALPINE GTA

YEARS MADE: From 1986.
ENGINE: V6, 2458 cc, single overhead camshaft, turbocharged, 200 bhp at 5700 rpm.
FRAME: Tubular backbone with integral fabricated subframes front and rear.
PERFORMANCE: 151 mph (243 km/h).
CURRENT PRICE GUIDELINE: * * *
Limited production and availability.
PROBLEMS TO LOOK FOR:
Mistreatment in secondhand examples.
CLASSIC CAR STATUS: Ready-made classic.

Passenger accommodation in the tiny A110 was always very limited and there was hardly any room for luggage, which confined sales, in general, to a young – and relatively impecunious – sector of the market (or the French police!). So Alpine introduced the larger A310, an early forerunner of the GTA Turbo, in 1971, with tiny rear seats which were meant to make it a competitor for the evergreen Porsche 911. Aside from the more modern styling, the A310 was basically similar to the earlier Berlinette except for the swing axle rear suspension which had been an Achilles heel for less-than-expert drivers. The A310's more conventional wishbone suspension was then transferred to the last of the A110s, some of which were produced with the very powerful 1800 cc engines from 1973.

However, the Alpines were rapidly becoming also-rans in top-line rallies as the new, purpose-built, Ferrari-engined Lancia Stratos (see page 104) began to make its mark.

As the last A110s were built in 1975, and as the A310 demonstrated that it was too bulky to take its place in competition, Alpine sales suffered badly in the wake of a worldwide energy crisis. The small firm from Dieppe was saved when it became a wholly-owned subsidiary of Renault.

One of Renault's first moves in 1976 was to adapt the A310 to receive its new and more powerful 2.7-litre V6 engine. It then enjoyed reasonable sales and left Renault to concentrate on turbocharging to capitalize on its pioneering work in Formula 1. The first turbocharged Renault-Alpine was to be a stubby, mid-engined two-seater which followed the same basic construction as the previous backbone-chassis cars, but was clothed in glass-fibre panels to resemble Renault's top-selling minicar, the 5. As such, the new Renault 5 Turbo, with up to 250 bhp and racing car suspension, became nearly as big a force in rallying as the Berlinette had been.

The A310 was then redesigned with a striking new low-drag body as the 160 bhp GTA in 1984 and, as the Renault 5 turbo became outdated in competition, development came full circle. The GTA was succeeded in 1986 by the latest GTA V6 Turbo with a full 200 bhp and so much pulling power that it doubled as both a sophisticated road car and a formidable competition machine.

SUMMARY

Plus points
High performance.
Attractive body.
Roomy cabin.

Minus points
Lack of luggage space.
Rear-engined lack of directional stability in sidewinds.

▽ ***The streamlined Renault-Alpine GTA V6 Turbo, a strong competitor for Porsche in the supercar stakes.***

ASTON MARTIN

Aston Martins have always been the English alternative to the most revered exotic cars in the world . . . Bugatti before the Second World War, Ferrari afterwards. In fact, it is hard to imagine a car more English in concept. The marque owed its existence to two former racing cyclists, mechanical engineer Robert Bamford and wealthy Old Etonian Lionel Martin. They progressed during the Edwardian era to classic trials and hill-climbs, using the small Singer sports cars which were popular at the time. Being relatively cheap machines, the Singers were far from perfect and Bamford, for one, was sure he could improve on them. He had considerable ability and soon built far superior, although necessarily more expensive, cars which took the name Aston Martin from Martin's exploits at the Aston Clinton hill-climb in Buckinghamshire. Soon the wealthy amateurs who took part in such events with their road cars were asking for replicas of Aston Martins. Bamford and Martin were happy to oblige, but, being English, they felt that it was not sporting to be too businesslike and make a profit from friends and the firm nearly folded in 1925.

It was saved by two 'high-society' enthusiasts, Lord Charnwood and the Hon. John Benson, who appointed an Italian-born managing director, Augustus Cesari Bertelli. He was not only a first-class manager, but also a gifted designer, excellent constructor and brilliant salesman, and he knew better than to change the essential character of the well-bred road cars. It seems that he was treated like a top-class horse trainer by the wealthy and often aristocratic customers who would breeze into Aston Martin's works, greeting him with, 'I say, Bertie, old boy . . .'

ASTON MARTIN ULSTER

YEARS MADE: 1934–5.
ENGINE: In-line four-cylinder, 1495 cc, single overhead camshaft, 80 bhp at 5250 rpm.
FRAME: Twin girder, cross-braced.
PERFORMANCE: 100 mph (161 km/h).
CURRENT PRICE GUIDELINE: **** Very rare, expect long wait for a good one.
PROBLEMS TO LOOK FOR: Check that components are original and not suffering from fatigue.
CLASSIC CAR STATUS: **** Likely to hold value.

The four-cylinder Aston Martins could, when required, put up a good show against the best Continental cars. Bertelli's greatest work, however, was the closest he came to an outright racing

▽ ***Aston Martin's greatest pre-war car, the Ulster, was designed as a road racer to match the finest products of rival company Bugatti.***

car. It featured an advanced dry-sump engine and was called the Ulster after the Ulster TT road race, where team cars had performed well. The Ulster was produced in small quantities between 1934 and 1935, but cost so much to build that Aston Martin had to return to less exotic cars when Bertelli retired in 1936.

SUMMARY

Plus points

Rarity value.

Best-performing of the pre-war Astons.

Numerous sporting events in which it can be driven.

Gilt-edged investment.

Minus points

Needs racing-car-style maintenance.

Parts supply good, but necessarily expensive.

Not so fast as some similar-price contemporaries.

△ ***Aston Martin's DB3S was given a variety of body styles: this coupé is one of the rarer versions.***

DB2

YEARS MADE: 1950–3.
ENGINE: In-line six-cylinder, 2580 cc, twin overhead camshafts, Standard form 105 bhp at 5000 rpm, Vantage 123 bhp at 5000 rpm.
FRAME: Tubular.
PERFORMANCE: standard 109 mph (175 km/h), Vantage 117 mph (188 km/h).
CURRENT PRICE GUIDELINE: *** Rare, but something of a bargain when ranged against Italian contemporaries.
PROBLEMS TO LOOK FOR: Neglect and general metal fatigue.
CLASSIC CAR STATUS: *** Purity of line ensures potentially higher rating than later, and more common, DB2/4.

Aston Martin continued to concentrate on commercially viable cars until the Second World War halted production. After the war the impetus was gone and the company found itself without enough capital. But once more it was saved by a wealthy patriot, tractor magnate David Brown, who also bought Lagonda. Brown intended running these companies almost as a hobby, but had the good fortune to find himself with an excellent new coil-sprung chassis from Aston Martin and an equally good twin-overhead-camshaft engine from Lagonda.

Initially, Aston Martin produced a few rather pretty touring cars – called the DB1 after the new owner's initials – to use up old parts before the Aston Martin's chassis was combined with the Lagonda's engine in 1950 to make the Aston Martin DB2. The name Lagonda, which had always been associated with bigger cars, was saved for larger touring versions.

◁ ***The Aston Martin DB2 (this is a prototype) emerged from amalgamation with Lagonda.***

SUMMARY

Plus points

Good value for an early-1950s grand touring car.

Excellent handling.

Numerous sporting events in which it can be driven.

Underrated investment.

Minus points

Costly to rebuild.

Tendency for metal fatigue in some castings.

DB3S

YEARS MADE: 1953–6..
ENGINE: In-line six-cylinder, 2922 cc–2992 cc according to specification, twin overhead camshafts, 182 bhp at 5500 rpm to 236 bhp at 6000 rpm, according to specification.
FRAME: Tubular.
PERFORMANCE: Circa 150 mph (241 km/h) according to specification.
CURRENT PRICE GUIDELINE: ****
Gilt-edged investment.
PROBLEMS TO LOOK FOR: Neglect and mistreatment, non-original restoration.
CLASSIC CAR STATUS: ****
Much-loved 1950s sports racer.

The DB2's performance in competition encouraged owner David Brown to run a team of purpose-built sports racing cars, the first of which was the DB3. This became the legendary DB3S in 1953 when more power was extracted from the Lagonda engine and it was clothed in a stunning, new, open two-seater body.

The DB3S was successful chiefly because it handled so well and was strong enough to take over the lead in long-distance races when more powerful opposition from Jaguar, Ferrari and Maserati faltered. But really it needed more power for outright victories and the priority became a new engine, which made its début in the DBR2 sports racing car.

SUMMARY

Plus points

Excellent performance.

Forgiving handling.

Practical competitor in numerous historic events.

Minus points

Some parts, such as twin-cam works cylinder heads, very expensive.

Needs racing-car maintenance and care with lightweight body.

▽ ***Italian coachbuilders Zagato specialized in ultra-lightweight bodywork, as fitted to this special competition version of Aston Martin's DB4GT.***

DB4GT ZAGATO

YEARS MADE: 1960–1.
ENGINE: In-line six-cylinder, 3749 cc, twin overhead camshafts, 314 bhp at 5500 rpm.
FRAME: Steel platform, alloy tubes for body.
PERFORMANCE: 152 mph (244 km/h).
CURRENT PRICE GUIDELINE: **** Gilt-edged investment.
PROBLEMS TO LOOK FOR: Non-original specification.
CLASSIC CAR STATUS: **** Highly rated 1960s competition car.

By the mid-1950s the market for very expensive sports racing cars was even smaller than before the war because they had become so specialized that they were no longer suitable for everyday use. And the customers who bought the touring cars, which basked in the racers' reflected glory, almost invariably wanted a more luxurious machine for what was still a high price. As a result, the DB2 had been developed up to a Mark III version with two small rear seats, and was beginning to show its age, so Aston Martin started designing a new grand touring car at the same time to capitalize on the sports racing cars' success.

This took the form in 1958 of the DB4, which had a massive steel floorpan carrying an aluminium body by the Italian firm Touring as stunning as that of the DB2 before it. And the performance, with new disc brakes and the far more powerful engine developed from that of the DBR2, made it a match again for the best that Ferrari could produce.

Like the DB2, the DB4 performed well on the track, particularly in short-wheelbase form with a full race engine. This model was known as the DB4GT, but it suffered against pure competition cars because – being based on a luxurious tourer – it carried too much weight. Meanwhile, the world sports car championship, which Aston Martin had won in 1959 with its sports racers, had been switched to GT cars. Aston Martin, naturally, tried to stay in the race by fitting special versions of the DB4GT with an ultra-lightweight Zagato body from 1961. These were eventually outclassed in international competition by the Ferrari 250GTO (see page 68) which was, in reality, a pure racing car. The DB4GT, however, remained the classic Aston Martin until the basic grand touring cars, in more luxurious, longer-wheelbase DB6 form by 1965, again began to suffer from relatively inadequate power because they had grown bigger and heavier.

SUMMARY

Plus points

Short production runs ensure ultimate exclusive appeal and value.

Reasonably good performance.

Good availability of mechanical spares.

Minus points

Not so rapid or nostalgic as a Ferrari 250GTO.

Agricultural gearchange.

△ ***The V8 Vantage, one of the fastest accelerating cars in the world.***

VANTAGE V8

YEARS MADE: 1977 to date.
ENGINE: V8, 5340 cc, four overhead camshafts, 485 bhp at 6200 rpm.
FRAME: Steel platform.
PERFORMANCE: 168 mph (270 km/h).
CURRENT PRICE GUIDELINE: New ****, secondhand **.
PROBLEMS TO LOOK FOR: Neglected maintenance.
CLASSIC CAR STATUS: *** Set to improve when model becomes obsolete.

Once more, Aston Martin was well on the way to producing an even stronger engine, of V8 formation rather than the earlier six cylinders. There were teething problems, however, and the six-cylinder engine had to be used initially in the new DBS grand tourer, which had a widened floorpan and totally revised, even more luxurious body. But when it received the new V8 in 1970, it became one of the world's fastest and most luxurious cars.

By this time, David Brown was ready to back down from what had been a very expensive hobby and, such are the economics of building exotic cars for a handful of customers, it had taken several saviours to keep Aston Martin afloat. Eventually, oil man Victor Gauntlett succeeded in grand style with the Aston Martin Vantage, boasting such a powerful engine that there is little on the road that can stay with it.

SUMMARY

Plus points
- Stunning performance.
- Good handling.
- Muscle-bound good looks.
- Excellent service available from factory.

Minus points
- Ageing design.
- Very heavy car.
- High price when new.
- High depreciation (although that can be an advantage to later purchasers!).

△ ***The Boat-Tailed Speedster is one of Auburn's best-loved products. Owner: The Auburn-Cord-Duesenberg Museum.***

AUBURN

Frank and Morris Eckhart produced their first car in 1900 and called it after the town where it was made: Auburn, Indiana.

The Auburn motor company continued in a fairly conservative way until 1924. At this point the firm was close to collapse, and it was bought by the greatest car salesman America has ever seen: Errett Lobban Cord.

AUBURN SPEEDSTER

YEARS MADE: 1934–5.
ENGINE: In-line eight-cylinder, 4585 cc, side valves, supercharged, 150 bhp at 3500 rpm.
FRAME: Twin girders, cross-braced.
PERFORMANCE: 103 mph (166 km/h).
CURRENT PRICE GUIDELINE: * * * * Rarity value dictates high prices.
PROBLEMS TO LOOK FOR: Mechanical deterioration, non-original parts.
CLASSIC CAR STATUS: * * * Tends to be over-priced compared to real exotics.

Cord, who had made $100,000 commission selling cars in California, took one look at Auburn's stock of 700 deadly dull cars, brightened them up with flashy paintwork and sold them almost overnight. His secret: he had pitched their appeal at hard-working young customers who had plenty of money, but who were considered too brash by the middle-aged salesmen that Auburn used to employ. Sales multiplied during the next five years as Cord made additional millions on the stock market, acquiring other companies, such as exotic car makers Duesenberg, and aero-engine manufacturers Lycoming.

Then the bubble burst. Car sales plunged following the Wall Street crash of 1929, but Cord's companies survived by slashing development costs. Inevitably, profits fell, too, and Cord decided in 1932 to take a gamble on producing fewer cars, but those which made more money because they sold at a higher price. The profits were high on every one of the few Cadillac, Packard and Pierce-Arrow V12 cars which could be sold, and Cord wanted one of these high-earners, too. He figured he could sell anything as long as it looked good, in spite of the depression.

So he extracted an instant V12 engine from Lycoming and installed it in a cheap Auburn chassis for 100 mph (160 km/h) performance. Even when fitted with rakish Speedster bodywork – from another Cord company desperate for work – the new Auburn could be sold at one-third of the price of rival models. Cord's gamble failed, however, as the people who could afford a V12 did not want a cheap car, and the people who wanted a cheap car did not want the expense of a V12 engine. He was in trouble over his share-dealing, too, so he emigrated to England in 1933 and left his car companies in the control of Duesenberg's Harold Ames.

Ames followed Cord's principles, revamping the Speedster with the help of Duesenberg's master stylist, Gordon Buehrig. He also replaced the hard-to-sell V12 engine with a thoroughly old-fashioned, but very cheap, Lycoming straight eight-cylinder unit.

Buehrig worked wonders with 100 redundant V12 Speedster bodies, chiefly by the cheap and easy solution of cutting off the back, so transforming the rear profile into what was to become a very trendy boat-tail. He also fitted new wings, safe in the knowledge that a stockpile of old wings could be sold cheaply to body repair shops. The styling transformation was completed by a shining new radiator, and gleaming chrome-plated exposed exhaust pipes like those on a Mercedes SSK (see page 118).

Another part of Cord's empire, the Columbia axle company, had fallen on hard times, so the Boat-tailed Speedster acquired a very cheap additional two-speed gearbox to supplement its original three ratios, thus achieving the technical distinction of becoming America's first six-speed car!

To the amazement of everybody at Auburn, the Boat-tailed Speedster then showed itself capable of cruising at 60 mph (97 km/h) in almost complete mechanical silence because the engine turned over so slowly in its very high sixth gear. It was a feat which Cadillac had failed to achieve despite massive investment over many years.

Such cars were not enough to save Auburn, however. But recently, now everyone has forgotten that the expensive looking Auburns were really only cheap, flashy cars, the Speedster has become one of the all-time classics . . . some say of the con-man's art!

SUMMARY

Plus points

Inspired styling.

Surprising performance.

Relative durability.

Minus points

None, providing you do not try to kid yourself that it is a Duesenberg!

AUDI

It has long been realized that four is the ideal number of wheels to have on a car, and, for more than 60 years, that brakes on all four wheels are the best way of stopping it. The same ideals apply to propelling a car, yet it was only recently that any manufacturer was able to produce a viable way of driving all four wheels. It was Audi – an upmarket division of the giant Volkswagen combine – which caused a sensation throughout the motoring world in 1980, when it announced its advanced four-wheel-drive, high-performance road car – the Quattro.

QUATTRO

YEARS MADE: From 1980.
ENGINE: In-line five-cylinder, 2144 cc, single overhead camshaft, turbocharged, 200 bhp at 5500 rpm.
FRAME: Integral.
PERFORMANCE: 137 mph (220 km/h).
CURRENT PRICE GUIDELINE: New * * *, secondhand * Heavy depreciation in common with top-of-the-range cars leaves secondhand Quattros dramatically undervalued.
PROBLEMS TO LOOK FOR: Neglected maintenance, misuse.
CLASSIC CAR STATUS: * * * Set to improve once the early models have become obsolete.

In the years before the Audi Quattro made its first appearance in 1980, all previous four-wheel-drive systems had foundered through high cost, weight, bulkiness, poor efficiency or unsatisfactory behaviour on the road.

Four-wheel drive became commercially viable for ordinary cars once shafts had been made with joints strong enough to enable the front wheels to move in any direction as they were driven, were steered, or absorbed road

▽ Audi's sensational Quattro Turbo Coupé set new standards with its pioneering four-wheel-drive system.

shock. Like most other motor manufacturers, Audi quickly established a range of front-wheel-drive vehicles which offered more room for passengers because they did not need the shaft from the rear-wheel-drive days that connected the engine at the front to the wheels at the back. This was especially important in the smallest cars, where space was at a premium.

For four-wheel drive to become a reality in a mass-produced car, all that was needed was a means of harnessing the rear wheels to the front without the harsh, noisy, heavy and expensive complication of a secondary gearbox, like that used on previous all-wheel-drive vehicles.

The answer came when Volkswagen and Mercedes became involved in a competition to design a new off-road vehicle for the German army as an alternative to the truck-like Land-Rover and its derivatives.

To the surprise of many observers, the Audi arm of Volkswagen won the contest in 1978 with a vehicle called the Iltis, using four-wheel drive for maximum traction and braking on slippery surfaces, and an equally surprising five-cylinder engine.

In essence, the Iltis system was far superior to its predecessors because it was based around two shafts, one running inside the other, which were lighter and quieter, occupied less space, and did not need so much power to operate as the previous profusion of cogs and spindles.

The incorporation of three differentials in the driveline – one in each axle and one in the gearbox – also helped to cut dramatically the increase in fuel consumption and tyre wear, which had always been a bugbear on four-wheel-drive vehicles.

The five-cylinder engine had already been introduced in the larger Audis. This unit proved to be a good compromise between the space-saving qualities of a conventional four-cylinder unit and the inherent smoothness of a six-cylinder.

It did not take much imagination to realize what such a compact and efficient pairing could do in a car, but the concept was still too unconventional for Volkswagen's management to accept it immediately.

However, one of their leading engineers, Walter Treser, was a keen rally driver and fitted a small Audi 80 saloon with the Iltis transmission to demonstrate how well it would handle on loosely surfaced roads. Once more, to everybody's surprise, it was not only good on the rough, but much better on smooth roads, too!

Treser needed only to demonstrate to Audi's chief salesmen that his four-wheel-drive prototype could climb the most difficult snow-covered mountain in Austria to win an audience with Volkswagen's top management.

In this case, Treser used the hoses of a factory fire engine to turn a local hillside into a greasy quagmire which only his prototype could climb, and finally got the go-ahead for production.

The system was first introduced on a new top model in the Audi range, a muscular coupé which could use a turbocharger now that it was possible to convey through all four wheels the extra power and torque which the front two would not have been able to handle on their own.

In fact the big turbo coupé – called the Quattro after its pioneering four-wheel-drive system – proved so fast and surefooted that it dominated world rallying for years. And the four-wheel-drive system was introduced as an option throughout the rest of Audi's range of predominantly medium- and large-size cars, where its advantages made up for the slightly bigger space it occupied.

The extent of Audi's achievement can be gauged from the fact that it took several years for the rest of the world's car companies to catch up.

SUMMARY

Plus points

- Extraordinary surefooted handling.
- Tremendous punch of turbocharged engine.
- Great strength.
- Roomy cabin.
- Surprisingly low fuel consumption.
- Excellent availability of parts.

Minus points

- Poor gear ratios.
- Awful ventilation.
- Expensive parts.

AUSTIN

The cheap and charming Austin Seven became Britain's most popular car before the Second World War and has lost little of this status since. It sold in large numbers because it was so cheap, at a time when many people had neither owned nor driven a car before. Despite its low cost, it was a very strong little car and lasted a long time, which meant that secondhand examples usually had many owners. As a result, for many people it was a first car in which they learned to drive.

AUSTIN SEVEN

YEARS MADE: 1922–31.
ENGINE: In-line four-cylinder, 696 cc, (from March 1923) 747 cc, side valves, 10 bhp at 2400 rpm.
FRAME: Twin girders, cross-braced.
PERFORMANCE: 42 mph (68 km/h).
CURRENT PRICE GUIDELINE: * May increase marginally as cars grow older. Some rare models, such as Ulster supercharged sports, much more valuable.
PROBLEMS TO LOOK FOR: Sheer old age and neglect, non-original parts.
CLASSIC CAR STATUS: ** Underrated, although Ulster is less so.

Part of the charm of being involved with an Austin Seven was getting to know some of its amazing quirks which made it easy to drive badly and difficult to drive well. Bad drivers gained great satisfaction from being able to handle it at all, and good drivers gained even more from overcoming its problems.

Most of the deficiencies sprang from a feature that was very common in the Austin Seven's day: an extremely flexible chassis. It was made from two small girders laid out like a capital A, with one leaf spring across the top to support the front axle and two extending from the bottom legs to link up with the back axle. This allowed all four wheels to twist about so much that an Austin Seven could clamber along unmade tracks – of which there were many when it was introduced in 1922 – with all the agility of a mountain goat. But this lack of rigidity also contributed to its disconcerting handling once it found its way back to the road.

As the solid front axle flapped about and wheels went over bumps, the body swayed from side to side, causing the car to enter into a series of gentle swerves. Most Austin Seven drivers, being new to motor cars, assumed such handling was normal, and – treating the car like a horse – reined it in by sawing the steering wheel from side to side in rhythm with the bumps. Luckily, this happened to be the best thing to do. But the effect was made all the more dramatic as one rear spring flattened and the other arched its back in sympathy when the car took a corner. The flattened spring then pushed one side of the axle backwards and the arching spring tugged the other side forwards, steering the back of the car in the opposite direction to the front. Again, the novice driver thought this was normal and corrected automatically, which caused the back springs to reverse their action and start the sequence all over again. Thus Austin Sevens cornered in a series of wild swerves, at the same time bouncing vigorously on their very flexible springs and chassis. They also had a clutch pedal with hardly any travel which, in less-than-expert hands, caused them to take off like kangaroos – but, again, new drivers thought this was quite normal!

Braking could be equally amusing, as the front axle twisted so much that the cables slackened, leaving most of the power applied to the back, with a sudden broadside as the result.

However, the man whose firm made the Seven, Herbert Austin, maintained that good brakes encouraged bad driving, and his word was accepted as gospel.

The Austin Seven also had a very small and economical engine with cylinders of a tiny diameter – or bore – because car taxation was based on that dimension at the time. As a result, the cylinders had to be very deep, which made for considerable pulling ability. This also meant that the engine did not produce much power, which, in turn, meant that it was very understressed and lasted a long time. The end result was that, despite its diminutive size, the Austin Seven was capable of hauling four full-sized adults, or the most amazing loads, very economically, although at a quite slow speed. This gentle pace was just as well, in view of the handling, and it ensured that few people came to grief in an Austin Seven. Indeed millions revered it as the source of their first experience of motoring.

△ ***The eccentric little Austin Seven taught so many people to drive that it became a national pet.***

Austin Sevens were made in numerous forms – from commercial vans to racing cars – until 1939.

SUMMARY

Plus points

Appealing nature.

Economy of operation and restoration.

Durability.

Good availability of parts, ideal for self-maintenance.

So small it is easy to store.

One of the most practical vintage cars.

Minus points

Antique handling and braking.

Low performance.

AUSTIN-HEALEY

Former competition driver Donald Healey created his own firm to build specialist sports cars using Riley engines in 1945, and he was one of the first Britons to realize the potential of the huge American market in the immediate post-war years. It is however, for his collaboration with the Austin division of BMC, which produced the much-loved Austin-Healey 100, the 3000 and the Sprite, that he is chiefly remembered.

AUSTIN-HEALEY 100

YEARS MADE: 1953–5.
ENGINE: In-line four-cylinder, 2600 cc, overhead valves, pushrods, 90 bhp at 4000 rpm.
FRAME: Integral.
PERFORMANCE: 107 mph (172 km/h).
CURRENT PRICE GUIDELINE: * *
Comparatively rare, good examples difficult to locate except in Californian-style climates.
PROBLEMS TO LOOK FOR: Extensive corrosion in damp climates, bleached out interior and electrical deterioration in popular Californian market, general mechanical wear and mistreatment. Some parts very difficult to obtain.
CLASSIC CAR STATUS: * * Very rare 100S model far more valuable.

After the Second World War, Healey's main problem was lack of capital to make his own mechanical components, so he was always on the lookout for cheap new sources of power. His biggest opportunity came when Austin's Atlantic coupé flopped in America and left the giant British company with a stockpile of redundant power trains. Healey promptly had a chassis made that could accommodate this running gear, along with a very attractive body from his stylist Gerry Coker. When Austin's chief, Leonard Lord, saw the prototype at the London Motor Show in 1952, he was so impressed that

▽ Donald Healey's 100 sports car, produced by Austin, became one of its most successful products.

he immediately came to an arrangement with Healey to mass-produce the car at his works in Longbridge, Birmingham. This royalty deal suited Healey well because he could produce only about five of the cars per week against the 200 that could be made at Longbridge. The cheap new car was then named the Austin-Healey 100, a particularly apt choice of title because it was the first 100 mph (161 km/h) car that the average person could afford.

In its earliest form, the Austin-Healey 100 was powered by the Atlantic's four-cylinder engine and represented such a bargain that it sold well, even though it had only a three-speed and overdrive gearbox. This deficiency was rectified in 1955 when the Austin A90 saloon, which shared the same running gear, acquired a four-speed gearbox that also found its way into the Austin-Healey. High-performance 100M and sports-racing 100S models were produced at the same time.

But when the big Austin saloon received a six-cylinder engine in 1956, the move was not so popular because the engine proved inferior in a revised Austin-Healey, known as the 100-Six. (The lustier early models then retrospectively adopted the designation 100/4.) Production was also transferred soon after to the factory at Abingdon, Oxfordshire, which produced one of Austin-Healey's chief rivals, the MGA (see page 128). The two cars were produced side by side in what was to become the biggest sports car factory in the world – in 1952 Austin had amalgamated with Morris, which produced M.G.s, to form the British Motor Corporation.

SUMMARY

Plus points

Spartan appeal.

Excellent performance offers good value for money.

Very durable once in good condition.

Minus points

Prone to extensive corrosion in damp climates.

Handling can be wild.

Lack of ground clearance.

Difficulty in obtaining some parts.

△ ***The first Austin-Healey Sprite, with its cheeky 'Frogeye' styling, has become an all-time classic.***

FROGEYE SPRITE

YEARS MADE: 1959–61.
ENGINE: In-line four-cylinder, 948 cc, overhead valves, pushrods, 43 bhp at 5000 rpm.
FRAME: Integral.
PERFORMANCE: 83 mph (134 km/h).
CURRENT PRICE GUIDELINE: * Undervalued in view of cheeky, practical charm.
PROBLEMS TO LOOK FOR: Extensive corrosion in damp climates, non-original parts, especially glass-fibre bonnet in place of metal, general wear and mistreatment.
CLASSIC CAR STATUS: * Likely to improve because of practicality and individual appearance.

Despite the amalgamation of Austin and Morris, there was still a lot of rivalry between the former competitors. Lord, head of the new company, having once been sacked by Morris, much preferred Austin-Healeys to M.G.s so, in 1957, he asked the still-independent Donald Healey to produce a new small sports car for him instead of approaching M.G., which was the traditional leader in this field. The result was the Austin-Healey Sprite of 1958 which, like the six-cylinder Big Healey, represented an outstanding bargain. Healey's only problem was that he had lost his stylist, Coker, and the appearance of the new small car – with headlamp fairings like a frog's eyes – caused a lot of amusement.

The 'Frogeye Sprite', as the smaller car was known, was eventually restyled along the more conventional lines planned for M.G.'s new sports car, the MGB (see page 129). The new Sprite made its début with a squared-off body, but few other changes, in 1961. It now bore either an Austin-Healey badge, or, with slightly different trim, M.G. decals with the name Midget.

In company with the Big Healey, the 'Spridget,', as the smaller car became popularly known, was updated mechanically from time to time.

But the Sprite vanished when BMC ended its royalty deal with the Healey family in 1970. The Midget carried on – with numerous modifications, including a Triumph engine from 1974 – until a plunge in the value of the US dollar in 1980 made it uneconomical to produce.

SUMMARY

Plus points

Spartan appeal.

Extremely durable, reliable and cheap to run once in good condition.

Good availability of cheap mechanical parts.

Minus points

Relatively low performance.

Scarcity of parts, such as very expensive one-piece metal bonnet as original.

3000 MARK III SERIES II

YEARS MADE: 1964–7.
ENGINE: In-line six-cylinder, 2912 cc, overhead valves, pushrods, 148 bhp at 5200 rpm.
FRAME: Integral.
PERFORMANCE: 120 mph (193 km/h).
CURRENT PRICE GUIDELINE: * * Far more readily available than earlier models.
PROBLEMS TO LOOK FOR: Extensive corrosion in damp climates, bleached-out interior and electrical deterioration in popular Californian market, general mechanical wear and mistreatment.
CLASSIC CAR STATUS: * * Brutal appeal puts it in front of many competitors.

The Big Healey, as the six-cylinder car was known, had been uprated from its 2.6-litre form to 3 litres as the 3000 in 1959, when, like the Frogeye, it became a top-seller for BMC.

Sales of the Big Healey were also boosted when the M.G. Competitions Department at Abingdon managed to develop it into one of the most formidable cars in international rallying, with drivers like Pat Moss winning such epics of the road as the Liège-Rome-Liège. The Healey family, meanwhile, continued to develop the 3000, along with the Frogeye Sprite, and Midgets which followed, for circuit races such as the Le Mans 24-hour.

The Big Healey continued to be modified, in line with changes in specification to the equivalent Austin Westminster saloon, but it eventually went out of production in 1967, squeezed out of its biggest market by new American environmental laws.

SUMMARY

Plus points

Excellent performance offers good value for money.

More comfortable than a 100/4.

Very durable once in good condition.

Revised rear suspension makes it the best of the 3-litres.

Reasonable availability of spares makes it easier to run than a four-cylinder.

Minus points

Prone to extensive corrosion in damp climates.

BENTLEY

Bentley has been one of the most emotive marques in British motoring history since its massive sports cars won the Le Mans 24-hour race five times between 1924 and 1930. The foundation for its success and the marque's prosperity was laid by Walter Owen Bentley, a former locomotive engineer who rated reliability above all other qualities.

W. O. Bentley had an eye for technological advances and made enough money from pioneering aluminium alloy aero-engine pistons during the First World War to establish his own car firm, Bentley Motors, in 1918. His first car featured a 3-litre, four-cylinder engine that was at least as advanced as that in any aeroplane – it had 16 valves and an overhead camshaft.

The massive chassis was also sufficiently competent to allow the new Bentley to win its first race at Brooklands in 1921, even though it had been designed purely as a touring car. It was found wanting, however, in what was to become the world's greatest sports car race, over 24 hours at Le Mans. The brakes – on the rear wheels only – which were considered adequate for normal use at the time, quickly wore out with hard use over such a distance. But when the 3-litre Bentley was fitted with brakes on all four wheels, it recorded the marque's first win at Le Mans in 1924.

▽ Last of the Big Healeys . . . the series II version of the Mark III.

BLOWER BENTLEY

YEARS MADE: 1929–30.
ENGINE: In line four cylinder, 4398 cc, single overhead camshaft, supercharged, 180 bhp at 3200 rpm.
FRAME: Twin girders, cross-braced.
PERFORMANCE: 125 mph (201 km/h).
CURRENT PRICE GUIDELINE: ***** Gilt-edged investment, rarest of the great Bentleys, every boy's dream of a blood-and-thunder sports car.
PROBLEMS TO LOOK FOR: Mechanical strain, non-originality.
CLASSIC CAR STATUS: ***** Worth waiting years for one of the few available.

Despite Bentley's 1924 Le Mans success, he could not survive by producing sports cars alone. The biggest potential market for heavy cars lay in luxurious saloons which needed an immensely powerful engine to haul them along at a decent speed. So Bentley produced an experimental six-cylinder, 4.5-litre engine in 1925 with the object of providing enough torque – or pulling power – to cope with the heaviest formal coachwork that any owner was likely to want on the established chassis. Bentley was confident that this engine would be a winner, but when a Rolls-Royce, with a prototype 7.7-litre engine, proved just as fast, he had to react by boring out his six-cylinder engine to 6.5 litres to stay ahead. This engine was also made to run more silently, like that of a Rolls-Royce, by a sophisticated new camshaft drive in place of the rumbling vertical shaft that had distinguished the earlier units. The development costs were enormous, however, and Bentley could only survive with backing from racing driver Woolf Barnato of the Kimberley diamond mining family.

The sports cars, although still huge machines, carried much lighter bodywork, so they could still win with smaller engines which consumed less fuel and did not need such a heavy chassis as the saloons. For these cars, Bentley designed a new four-cylinder engine of 4.5 litres – called the 4½-litre in those days – to win again at Le Mans in 1927.

A further victory followed in 1928, although the race was a close-run thing with an American Stutz pressing Barnato's car hard. This allowed the diamond miner to persuade the conservative W. O. Bentley to risk reliability by fitting the six-cylinder engine in the four-cylinder chassis. The gamble paid off, for they won in 1929 and 1930, despite intense opposition from Mercedes.

Works driver Henry Birkin had doubted, however, that the heavy-

△ ***The Blower Bentley became one of the marque's most romantic models. Owner: Nigel Dawes Collection.***

△ ***The superb 8-litre saloon was such a threat to Rolls-Royce that it bought up Bentley Motors.***

weight Speed Six model would be fast enough for the 1929 race, and he therefore followed an independent course. He obtained sponsorship from a wealthy patriot, Miss Dorothy Paget, to produce a team of lighter supercharged 4½-litre Bentleys, despite W. O. Bentley's objections that they would not be sufficiently reliable to last 24 hours. Birkin pointed out that rivals, such as Alfa Romeo, had made the supercharger a winning proposition, and eventually persuaded Bentley – against his better judgment – to produce 50 of these cars to qualify the model for Le Mans. The 'Blower Bentley', as it was called, proved so fast that it became Britain's most charismatic sports car, although it lacked the reliability to win any significant races and cost so much to make at a time of economic depression that Bentley Motors went into receivership.

SUMMARY

Plus points

Rarity value.

Built like a locomotive.

Wonderful nostalgia appeal.

Numerous events in which it can compete.

Minus points

Unwise to extend engine too far.

Mechanical restoration costs are in proportion to its very high value.

8-LITRE

YEAR MADE: 1930.
ENGINE: In-line six-cylinder, 7983 cc, single overhead camshaft, 220 bhp at 3200 rpm.
FRAME: Twin girders, cross-braced.
PERFORMANCE: 101 mph (163 km/h).
CURRENT PRICE GUIDELINE: **** Gilt-edged investment, biggest of the Bentleys.
PROBLEMS TO LOOK FOR: Non-originality, worn out components.
CLASSIC CAR STATUS: ***** As good as a Blower Bentley and more reliable, frequently held back by unadventurous original coachwork.

The firm had just produced a magnificent 8-litre saloon capable of more than 100 mph (161 km/h), that was a challenge to other exotics. In keeping with W. O. Bentley's belief in power without stress, its engine was a larger version of the superb 6½-litre six, producing 220 bhp and tremendous torque. It was built like a steam engine and pulled like one, too.

The magnificent 8-litre also featured a new four-speed gearbox and vacuum servo-assisted brakes at a price that brought it into direct confrontation with the Phantom II (see page 150), so it was hardly surprising that, with an encouraging order book, Bentley was promptly bought up by Rolls-Royce when the company ran into financial trouble. This spelled the end of the racing years and the Bentleys which followed were simply sporting versions of the more sedate 20–25 hp Rolls-Royce.

The equally serene 4.25-litre Bentley stemmed from the 8-litre saloon to take the marque into the post-Second World War market with a standard steel-bodied saloon, leaving Rolls-Royce unchallenged in a declining market for cars with bespoke coachwork.

SUMMARY

Plus points

Rarity value.

Built like a locomotive.

Sheer inherent reliability.

Minus points

Not so glamorous as a 'Blower Bentley'.

Bodywork often dowdy but still costs a fortune to restore.

R TYPE CONTINENTAL

YEARS MADE: 1952–5.
ENGINE: In-line six-cylinder, 4566 cc, overhead inlet, side exhaust valves, power output not revealed, probably about 140 bhp.
FRAME: Twin girders, cross-braced.
PERFORMANCE: 117 mph (188 km/h).
CURRENT PRICE GUIDELINE: * * *
Relatively rare model, set to appreciate dramatically. Deserves to be the most valuable post-war Rolls-Royce or Bentley.
PROBLEMS TO LOOK FOR: Terminal rust, non-originality, mistreatment of coachwork.
CLASSIC CAR STATUS: * * * *
Dramatically undervalued.

By 1951, the long-serving Bentley and Rolls-Royce engine had been uprated to 4.5 litres and H. I. F. Evernden, project engineer at Rolls-Royce, was given enough rein by a very conservative management to produce a special lightweight version. He sold the project to the management on the premise that he could produce a car which 'not only looked beautiful, but possessed a high maximum speed coupled with excellent handling and roadability'. The management gave way grudgingly, chiefly because of increasing criticism of its adoption for Bentley of the slogan 'The Silent Sports Car'. Many potential customers pointed out that, although the Bentleys built by Rolls-Royce at Crewe since 1931 were silent, they were certainly not sports cars like those built by W. O. Bentley at Cricklewood. They were, quite simply, Rolls-Royces with a different radiator and badges. Evernden's case received solid backing from Bentley's agent in Paris, Walter Sleator, who was responsible for a prototype of similar conception before the war. Sleator pointed out that such a model would sell well to his European clients, who had the advantage of miles of fast open road that were not available in Britain at the time.

This car, brilliantly styled by another Rolls-Royce employee, J. P. Blatchley, and produced by coachbuilders H. J. Mulliner, became the sensational Bentley Continental – so called because of its intended market. It proved the fastest saloon car in the world in 1952, and sold well until the Rolls-Royce on which it was based was redesigned in 1955 as the Silver Cloud with new bodywork that was more attractive than the standard line fitted earlier.

Bentley variants of this heavier new S-type Rolls-Royce continued to be produced, and the name Continental was retained for special versions until 1966 when a new series of 'Silver Spirit' cars was introduced.

SUMMARY

Plus points

Superb styling.

Beautifully built.

Marvellously relaxed, usable performance.

Good availability of spares.

Minus points

Restoration costs high, especially of aluminium body and interior.

▽ ***Bentley's R type Continental became not only the world's fastest saloon car, but also one of the most luxurious.***

△ ***The turbocharged Mulsanne has marked the re-emergence of Bentley as a separate marque, distinct from Rolls-Royce.***

MULSANNE TURBO

YEARS MADE: From 1982.
ENGINE: V8, 6750 cc, overhead valves, pushrods, power output not revealed, probably about 380 bhp.
FRAME: Integral.
PERFORMANCE: 133 mph (214 km/h).
CURRENT PRICE GUIDELINE: * * * *
Good value by super-saloon standards.
PROBLEMS TO LOOK FOR:
Mistreatment.
CLASSIC CAR STATUS: * * * Flawed by hard ride.

After the Silver Spirit, special-bodied Bentleys continued to be made on the T-type Rolls-Royce chassis, but they differed little from the basic Rolls-Royce and sold in very small numbers until, in 1982, a younger management introduced a much faster turbocharged model called the Mulsanne (after the famous straight at Le Mans) to recapture the glories of the 'Blower Bentley' which had triumphed in that famous race 50 years before.

It was a remarkable turnaround for the marque, as only a year or so before Rolls-Royce had considered dropping the name Bentley altogether. At that time, Bentley-badged variants of Rolls-Royce models accounted for only 5 per cent of total sales, but the resurgence of interest in Bentley created by the Mulsanne Turbo R led to new models in 1984 that moved the marque further away from its Rolls-Royce equivalents.

The first of these new cars was the Bentley Eight, using a similar specification to the normally aspirated Mulsanne, but with firmer and more sporting suspension, a bright mesh grille and a lower price. At the same time, the name Continental was revived for the first time since 1966 on a replacement for the Corniche featuring more modern detail styling. The result was a range that now provided more than 40 per cent of the parent company's sales.

SUMMARY

Plus points

Superb build quality.

Reliability.

Stunning performance for bulk.

Surprising agility with later sporting suspension package.

Minus points

Sheer size for a sporting saloon.

Very thirsty and noisy for a Rolls-Royce-inspired vehicle.

BMW

BMW has almost invariably produced cars to the high standards expected of a company which started as an aero-engine maker in 1916. By 1922 the company's name had been changed to Bayerische Motoren-Werke, hence the initials BMW, as its production base in Munich expanded to include engines for boats, trucks and motor cycles, with the first complete motor cycle appearing a year later. When BMW turned its attention to car production, it was never to look back, apart from a hiccup after the Second World War.

BMW 328

YEARS MADE: 1936–40.
ENGINE: In-line six-cylinder, 1971 cc, overhead valves, side camshaft, 80 bhp at 5000 rpm.
FRAME: Tubular.
PERFORMANCE: 93 mph (150 km/h).
CURRENT PRICE GUIDELINE: * * *
Relatively rare, availability scarce.
PROBLEMS TO LOOK FOR: Wear and tear associated with old age, non-originality.
CLASSIC CAR STATUS: * * * *
Exceptionally practical high performer.

BMW's magnificent motorcycles, produced from 1923, were especially suited for use with a sidecar, but it soon became apparent that the future of such combinations would be decidedly limited by the success in England of the Austin Seven (page 26), which cost little more and offered all the advantages of a four-wheeled car. Surprisingly, other car manufacturers were slow to copy the Austin Seven, and BMW became the first German manufacturer to cater for a glaring gap in its home market by producing the British car under licence as the BMW-Dixi at a new factory at Eisenach.

The relatively primitive engineering of the British car was hardly up to BMW's standards, however, and the Dixi soon began to change, acquiring a far more advanced tubular backbone chassis and independent suspension by 1932. Six-cylinder engines were introduced the following year, with triple-carburettor sports models which were able to compete with the best rally cars. By 1935, the beautifully built BMWs were being imported into Britain to be marketed under the old-established Frazer Nash badge.

Although they were relatively expensive, the quality and performance of these BMW saloons and sports cars were such that they sold well. By far the most successful of these cars was the 328 sports model which handled exceptionally well and boasted 80 bhp from an ingenious engine featuring hemispherical combustion chambers.

SUMMARY

Plus points
Excellent performance and durability combined with superb handling.

One of the most desirable small pre-war sports cars.

Minus points
Scarcity of many parts makes them expensive, although durability compensates.

▽ The pre-war 328 was the first great sports car produced by the Bayerische Motoren-Werke, known now as BMW.

3.0CSL BATMOBILE

YEARS MADE: 1973–5.
ENGINE: In-line six-cylinder, 3153 cc, single overhead camshaft, 206 bhp at 6000 rpm.
FRAME: Integral.
PERFORMANCE: 137 mph (220 km/h).
CURRENT PRICE GUIDELINE: * * Set to appreciate as corrosion decimates the survivors.
PROBLEMS TO LOOK FOR: Rampant corrosion in bodyshell, high-mileage mechanical wear, non-originality, many Batmobile clones built from basic 3.0CSL model.
CLASSIC CAR STATUS: * * * Likely to improve with increase in rarity of genuine examples.

BMW was badly hit by the Second World War, which left its main car factories, at Eisenach and Berlin, behind the Iron Curtain. In the early post-war days the motor cycles made in Munich therefore had to take priority. The first cars made by the re-formed company in Munich did not appear until 1951, and used an engine based on the pre-war six-cylinder unit. This Type 501 launched a line of prestige cars, some with V8 engines, which took BMW to the verge of bankruptcy. The company tried to counter this trend by producing under licence a very economical 'bubble car' called the Isetta, which developed in 1960 into its own 700 four-wheeler, powered, not surprisingly, by a motor-cycle engine. Then the local Bavarian government came to BMW's rescue with a financial package which allowed it to introduce a 'proper small car', the 1500, with a new four-cylinder engine, in 1962. The situation improved considerably for BMW and the company once more became a successful car manufacturer.

As the four-cylinder cars provided the solid backbone of volume sales, BMW was again able to think about producing larger luxury cars. The first of these in 1965 was a high-performance coupé based on the contemporary 1600 cc four-cylinder car's floorpan with a six-cylinder version of its engine. In BMW tradition, these coupés had developed into quite a pleasant touring car, the 2800CS, by the time that Ford, in 1969, began promoting its new coupé the Capri, on the race track.

BMW, which regarded Ford – with factories in Cologne – as a fellow German manufacturer, was aghast. Previously, it had always stuck to a nationalistic cartel system in which Porsche made sports cars, Mercedes luxury saloons, BMW high-quality touring cars, Ford of Germany and Opel cheaper saloons, and Volkswagen the Beetle (see page 158) – which was completely classless. The furore which followed, as BMW turned its big coupés into competition cars and Ford fought back, resulted in some of the greatest production car racing ever seen in the European Touring Car championship. Eventually BMW came out on top with an ultra-lightweight version of its 3.0CSL coupé which sported such dramatic aerodynamic wings that it

became known as the 'Batmobile', taking its name from a popular TV character's space-age car.

SUMMARY

Plus points
Excellent performance and comfort for four.

Batmobile wing set improves road grip.

Nostalgia for great touring car races.

Minus points
Very rust-prone.

Tail-happy handling even with wing set.

Expensive parts.

Many bogus examples to trap the unwary.

◁ ***BMW produced their 3.0CSL to qualify their big coupé for racing and then added a futuristic 'Batmobile' wing set that transformed its roadholding.***

M1

YEARS MADE: 1978–80.
ENGINE: In-line six-cylinder, 3453 cc, twin overhead camshafts, 277 bhp at 6500 rpm.
FRAME: Tubular.
PERFORMANCE: 162 mph (261 km/h).
CURRENT PRICE GUIDELINE: * * *
Gilt-edged investment.
PROBLEMS TO LOOK FOR:
Mistreatment, accident damage.
CLASSIC CAR STATUS: * * * * Set to improve as cars become older, especially in those built for competition.

The fierce competition stimulated by the shattering of Germany's cartel system consumed huge amounts of money and engineering time, so BMW and Ford were forced to scale down their activities when the world's first energy crisis hit sales badly in 1974. BMW was reluctant to give up a hard-won reputation on the race track, however, and collaborated with the Italian supercar firm, Lamborghini, to produce in 1978 a stunning new competition sports car called the M1. The eventual intention was to beat established leaders Porsche, but the cost of reducing the weight of this mid-engined supercar was too high and the M1 was never able to mount a serious challenge in either Group Four sports car racing – for which 400 had to be built to quality – or the far more exclusive Group Five racing in which a car with the silhouette of an M1 could have used as much as 850 bhp from a turbocharged engine. And so the M1 remained as 'just' a glorious road car, with BMW switching its plans for expansion into the Mercedes market for luxury saloons.

△ ***BMW joined with the Italian supercar-maker, Lamborghini, to produce the stunning M1 coupé.***

SUMMARY

Plus points
Tremendous usable performance.

Superb handling.

Excellent build quality for a supercar.

First-class value for money.

Minus points
Bulky.

Restricted rearward vision.

BRISTOL

Britain's aircraft manufacturers awaited the end of the Second World War with such trepidation that one in particular, Bristol, decided to produce cars as a useful sideline. It reasoned that this had always been profitable for one of its major competitors, Rolls-Royce, so it might occupy a workforce at least as skilled during drops in demand for aeroplanes. And Bristol's gamble was to pay off, the marque earning a reputation for top-quality workmanship in the high-performance luxury car market.

BRISTOL 401

YEARS MADE: 1948–53.
ENGINE: In-line six-cylinder, 1971 cc, overhead valves, pushrods, 85 bhp at 4500 rpm.
FRAME: Tubular.
PERFORMANCE: 99 mph (159 km/h).
CURRENT PRICE GUIDELINE: * * Dramatically undervalued, set to appreciate fast.
PROBLEMS TO LOOK FOR: Sheer old age, but fewer problems than normal because car was so well made.
CLASSIC CAR STATUS: * * * Underrated, set to appreciate when quality and originality of design—for a Bristol—is more generally recognized.

The main problem for Bristol in its early car-producing days was that it could not simply resurrect its pre-war designs, as Rolls-Royce and most other manufacturers were bound to do, and there was not enough time to design a new car from scratch if it was to swing straight into car production once the war was over.

Fortunately, however, it had taken on in 1940 a director called H. J. Aldington (whose company had manufactured Frazer Nash sports cars before the war, see page 76), chiefly because of his close links with aero-engine manufacturers BMW. So, when Bristol decided to plunge into car manufacture, he was the ideal person to give advice. Aldington suggested that Bristol should take over where BMW had left off before the war, particularly as the German firm was not in an immediate position to start making cars again (see page 36). Part of the deal was that, if Bristol manufactured BMW's 2-litre engine, Aldington could then install it in his own cars.

The idea was also appealing to Bristol because the pre-war BMW cars were of exceptionally good design, and it was anxious not to be associated with an inferior product which would tarnish an image for top-quality engineering, so vital in the aeronautics field. In fact the company pursued this policy to extremes, producing a car that was as good as a Rolls-Royce.

In essence, Bristol took the best of the BMW designs by incorporating the chassis of the latest 326 tourer (which had more sophisticated suspension than the 328 sports car), the latest 328 engine (which had powered 1940 Mille Miglia race-winning versions), and developing the bodywork of the 327 sporting saloon, because Frazer Nash was going to look after the sports car market. In the event Bristol made a brilliant job of combining these diverse elements into its Type 400 saloon of 1947. The overall quality of this car was helped by the fact that at the time Bristol was one of the few manufacturers to appreciate the importance of aerodynamics in body design and to have access to a wind tunnel.

The 400 stayed in small-scale production until 1949, by which time Bristol had been able to develop a fully streamlined body of its own. This car, the 401, was inspired by the Italian coachbuilder Touring – responsible at the same time for the classical Ferrari barchetta body (see page 65) – and built on its exceptionally light *superleggera* principles using alloy panels welded to a spider's web of small tubes. A Type 402 drophead coupé, of which only 24 were made, was introduced soon after, but the 401 – which stayed in production until 1953 – remains the most popualr model.

The BMW-inspired Bristol engine was, meanwhile, proving popular in not only Frazer Nashes, but other sports racing cars too, and had even been uprated to 125 bhp for the Cooper-Bristol Formula 2 racing car which challenged the might of Ferrari. With such a high-performance image being gained by their engine, the Bristol saloons seemed rather slow, so a 100 bhp version was fitted to the 401 which, with further improvements, became the 403 in 1953. At the same time, a limited run of short-wheelbase 404 sporting saloons was offered with the option of either 105 bhp or the full 125 bhp. As an alternative, from 1954, Bristol produced its first four-door car, the 405, before

△ ***After the war, the Bristol aircraft makers expanded into motor car production with the 401.***

introducing the 406 with a more flexible 2.2-litre engine and restyled body.

This formed the basis for the 407 in 1962 which had far more power and torque from an American Chrysler V8 engine, and the basic design continued with minor changes through the Types 408, 409 and 410 to three separate series of the 411 by 1978. This car was then replaced with a far more streamlined model, the 603, and followed by another convertible, the 412, and a turbocharged Beaufighter, to form the basis of today's limited-production but still top-quality range.

SUMMARY

Plus points

Superb build quality.

Advanced design of body.

Durability.

Relatively high performance.

Forgiving road manners.

Minus points

Cost of restoring *superleggera* body.

Expensive mechanical components.

BUGATTI

For every person who claims that Ettore Bugatti made the greatest sports cars in the world, there will always be others who put forward marques like Alfa Romeo, Bentley, Ferrari, Jaguar and Mercedes Benz for that honour. Everybody is entitled to an opinion, and Bugatti – Le Patron *– was certainly entitled to claim that Bentley made the fastest 'lorries' in the world, for his own cars invariably displayed far daintier and more elegant engineering.*

It suited Bugatti in more ways than one to produce cars with extremely finely honed lines: not only was he the product of an artistic Italian family, but the factory which he made his feudal empire relied on clean profiles to avoid too many complex machining processes.

Bugatti settled in the Strasbourg area when he was a young man designing cars for other people. But always he wanted to work for himself and eventually he found a backer to help him produce his own Bugatti car in a disused dye works at Molsheim. He called his first production Bugatti the Type 13, because he could lay claim to having designed 12 cars before it. Although all Bugattis bear type numbers, the sequence can be confusing because it relates to the order in which the models – or engines – were designed, rather than produced.

The Type 13 sold well between 1909 and 1914 and continued to do so after the First World War. Similar Types 15, 17, 22 and 23 produced until 1926 featured minor technical changes, although the designation Type 13 was reserved for the competition versions, no matter how they changed. With an advanced 1.5-litre, 16-valve, four-cylinder engine, the Type 13 won many races, including one at Brescia in 1921 that gave the competition model its name: Brescia Bugatti.

TYPE 37A

YEARS MADE: 1926–30.
ENGINE: In-line four-cylinder, 1496 cc, single overhead camshaft, 90 bhp at 4000 rpm.
FRAME: Twin girders, cross-braced.
PERFORMANCE: 100 mph (161 km/h).
CURRENT PRICE GUIDELINE: *****
Like all Bugattis, very expensive and unlikely ever to lose value.
PROBLEMS TO LOOK FOR: Non-originality, wear and tear of old age.
CLASSIC CAR STATUS: *****
Gilt-edged investment.

Following his early successes, Bugatti ventured into larger cars, introducing the Type 30 as his first eight-cylinder production model in 1922. This was a touring model which was quickly followed by one of the most beautiful racing cars ever built, the 2-litre Type 35, which spawned a whole series of classic Bugattis, including the Type 37A. The original Type 35 boasted a full roller-bearing crankshaft together with a brand new chassis, a supremely elegant body and distinctive cast-alloy wheels.

In a manner which became typical of Bugatti, it displayed superb roadholding without resorting to the complexities of independent front suspension by virtue of having stiff half-elliptic springs at the front and more flexible quarter-elliptics at the back, often supplemented by very expensive de Ramm shock absorbers. Ingenious design was never better illustrated than by the way the lightweight frame was cleverly stiffened by the engine's crankcase, and unsprung weight was saved by piercing the front axle so that the springs could pass through it. There were teething troubles in the beginning, but by 1926 the Type 35 had won the world championship. More than any other, this machine, produced by an Italian on the borders of Germany, led to Bugatti being adopted as the French national marque.

Bugatti realized, however, that he could not rest on his laurels; he would need more power to stay ahead because rivals such as Alfa Romeo were developing ever more powerful engines. So he followed the fashion of the 1920s and fitted a supercharger to produce the Type 35C, the Type 35B

△ ***The beautiful supercharged Bugatti Type 37A, derived from the legendary 2-litre Type 35 racing car, was one of the fastest sports cars of the 1920s. Owner: G. Perfect.***

with a 2.3-litre engine, and the Type 39 with a 1.5-litre power unit. A detuned eight-cylinder engine was fitted in the Type 35A touring car before a plain-bearing four-cylinder version of the straight-eight was fitted to the Grand Prix chassis to make the Type 37, which duly became one of the fastest sports cars on the road as the supercharged Type 37A.

Yet more combinations were to follow by 1929: the Type 38 touring car replaced the Type 30, then was in turn replaced by the 3-litre eight-cylinder Type 44, by a Type 40 using the Type 37's four-cylinder engine, and a Type 43 adopting Type 35B power to become the fastest sports car in the world.

SUMMARY

Plus points

One of the most beautiful Bugattis ever made.

Four-cylinder plain-bearing engine relatively durable, coil ignition simple and easy to maintain.

Can be driven on the road with great gusto.

Numerous events in which it can compete.

Minus points

Restoration costs can be very high, despite practical nature of engine.

TYPE 41

YEARS MADE: 1929–32.
ENGINE: In-line eight-cylinder, 12,763 cc, single overhead camshaft, 275 bhp at 1700 rpm.
FRAME: Twin girders, cross-braced.
PERFORMANCE: 100 mph (161 km/h).
CURRENT PRICE GUIDELINE: ***** +
Only six made, almost priceless.
PROBLEMS TO LOOK FOR: Old age.
CLASSIC CAR STATUS: ***** The ultimate investment.

After the spectacular Type 43 came Bugatti's most ambitious car, of which he had been dreaming since 1913. Its magnificent power unit was to be based on an aero-engine which Bugatti had drawn for the French government in 1923. Initially this straight eight-

▷ ***Bugatti's Type 55 offered scintillating road performance.***

cylinder unit was going to have a capacity of 14.7 litres, but it weighed so much that even Bugatti had to compromise and scale it down to a mere 12.7 litres. The finished car was still one of the biggest ever made. The reason for building everything on such a massive scale was that the chassis would have to haul the heaviest and most opulent coachwork in the world. This car was for kings and Bugatti called it the Royale. Only six were made and just three were sold, because even royalty could ill afford such extravagance during the Depression. Eventually its magnificent engine found a more mundane use in French trains.

SUMMARY

Plus points
Rarity value.

Sheer size offers unparalleled value for money.

Engine proved durable in rail cars.

Minus points
As vast and as difficult to manoeuvre as a rail car.

TYPE 55

YEARS MADE: 1932–5.
ENGINE: In-line eight-cylinder, 2262 cc, twin overhead camshafts, 130 bhp at 4500 rpm.
FRAME: Twin girders, cross-braced.
PERFORMANCE: 112 mph (180 km/h).
CURRENT PRICE GUIDELINE: *****
Gilt-edged investment.
PROBLEMS TO LOOK FOR: Non-originality, old age.
CLASSIC CAR STATUS: ***** One of the most beautiful Bugattis.

Although chastened by the financial disaster that the Royale had proved to be, Bugatti went on to produce more superb cars, mixed with a few oddities, and – typically – fell on his feet with another successful design. This was the 5.3-litre Type 46 – essentially a mini-Royale – which made a brilliant début and restored the Bugatti factory to financial health by selling strongly.

Two rather impractical 16-cylinder racing cars were made at the same time – the Types 45 and 47 – with another model of which only one was produced – the four-wheel-drive Type 53 hill-climb car. Then the Type 46 became the Type 49 with a twin-overhead-camshaft engine that was even too powerful for the safety of a Type 54 racing version!

But Bugatti was on the right track, retaining the twin-overhead-camshaft engine design for the Type 51. This

▽ ***Bugatti's massive Type 41. Owner: Musée National de l'Automobile.***

updated Type 35 Grand Prix car then became one of the most successful Bugattis ever, once more restoring the fortunes of the factory and its owner!

In 1933, the Type 51 was uprated as the Type 59, to become the fastest racing Bugatti of all. In the mean time the Type 55, a sports version of the Type 51, became one of the most desirable Bugattis ever made, offering a scintillating performance on the road.

SUMMARY

Plus points

Superb performance and handling.

Flowing lines.

One of the most pleasant Bugattis to drive.

Minus points

Exotic machinery needs very high standards of maintenance.

Potentially very costly.

TYPE 57SC ATLANTIC

YEARS MADE: 1936–8.
ENGINE: In-line eight-cylinder, 3257 cc, twin overhead camshafts, supercharged, 200 bhp at 4500 rpm.
FRAME: Twin girders, cross-braced.
PERFORMANCE: 113 mph (183 km/h).
CURRENT PRICE GUIDELINE: ***** + In the same league as the Royale.
PROBLEMS TO LOOK FOR: Old age.
CLASSIC CAR STATUS: ***** The definitive exotic car.

The touring equivalent of the Type 59 (see above), the Type 57 – produced from 1934 to 1939 – rose to even greater heights, especially in supercharged form as the Type 57C, and as the sporting Type 57S.

But the most spectacular Bugatti of all was the Type 57SC (sports/supercharged) coupé when it was fitted with fixed-head Atlantic bodywork. Only three of these beautiful cars were built between 1936 and 1938, but they will never be forgotten.

△ The Type 57SC Atlantic, with its riveted-spine bodywork and superb supercharged performance, is the most spectacular Bugatti ever made. Owner: Tom Perkins, California.

Ettore Bugatti survived the Second World War, but *Le Patron* and his factory were a spent force. He died in 1947 and only a few more cars were produced, none of them great. Those produced before the war, however, were not only the creations of a great artist and of incredibly skilled workers, but also superbly rewarding machines to drive.

SUMMARY

Plus points

Fantastic bodywork.

Superb performance.

Minus points

Bodywork restoration costs likely to be even greater than mechanics.

CADILLAC

From inauspicious beginnings in 1903, Cadillac rapidly became the standard-bearer for the world's biggest car manufacturer once it had been absorbed by the American General Motors group in 1909. Cadillac's position was further reinforced when General Motors was taken over by ace salesman, Alfred P. Sloan, in 1923, and Cadillac cars went on to become the country's greatest status symbol.

CADILLAC 452

YEARS MADE: 1930–6.
ENGINE: V16, 7353 cc, overhead valves, pushrods, 165 bhp at 3400 rpm.
FRAME: Twin girders, cross-braced.
PERFORMANCE: 84–100 mph (135-161 km/h), depending on body style and axle ratio.
PROBLEMS TO LOOK FOR: Non-originality, general old age.
CLASSIC CAR STATUS: **** It will stay on a consistent plane.

Owner A. P. Sloan's philosophy for selling cars to a fickle public was to provide enough different models – and makes – to accommodate everybody's needs 'from cradle to grave'. Thus, car ownership could progress from a starting point with General Motors' cheap Chevrolet, through Pontiac, Oldsmobile, Buick and La Salle to the top of the line: Cadillac. The theory was that children born into a family at any time during this progression of ownership would be indoctrinated with the General Motors' creed. According to Sloan's theories, this made even the most expensive limited-production Cadillacs just as important as the volume-production Chevrolets.

He took immediate action, therefore, when, in the 1920s, Cadillac lost its lead in large-car sales to Packard, which had followed the European trend towards V12 luxury cars. Sloan ordered Cadillac to produce a new car that was faster, more powerful and more refined than anything else in the showrooms. And it was no good simply enlarging the famous V8 engine to achieve this object, or producing a rival V12; the new car must have more cylinders!

△ ***Cadillac's opulent 452, with its 16-cylinder engine, is one of the most exotic cars ever. Owner: Paul Shinnerer.***

Sixteen was the next step up, so that's what the General Motors engineers produced; a wonderfully smooth unit featuring the world's first hydraulic tappets, which also made it uncannily silent.

Sales of the new Cadillac – called the 452 after its American-style capacity in cubic inches – were quite good in its first year, 1930. But they soon fell off because so few people wanted, or could afford, such an opulent motor car. Sloan, however, would not be moved, and Cadillac kept on producing a V16 until the Second World War really took hold in 1940, although a V12 version of General Motors' top car proved far more popular.

SUMMARY

Plus points
- Sixteen cylinders guarantees immortality.
- Excellent build quality.
- Practical durability.
- Easy to drive.
- Very comfortable.

Minus points
- Restoration costs as huge as the car.

ELDORADO

YEAR MADE: 1953.
ENGINE: V8, 5394 cc, overhead valves, pushrods, 160 bhp at 3800 rpm.
FRAME: Twin girders, cross-braced.
PERFORMANCE: 116 mph (187 km/h).
CURRENT PRICE GUIDELINE: ** Set to appreciate as a classic dinosaur.
PROBLEMS TO LOOK FOR: Non-originality, corrosion except in Californian style climates where trim and electrical deterioration is common.
CLASSIC CAR STATUS: *** Likely to improve as car grows older.

Styling has always played a vital part in struggles for car sales and Cadillac began to leave its rival Packard behind when, in 1941, it broke away from the traditional upright radiator grille, adopting a horizontal 'eggcrate' grille and fins on the tail which appealed to a younger generation. Packard, meanwhile, was hampered by its stodgy, traditional styling.

By 1950, Cadillac – resolutely back to a large-capacity V8 engine that was one of the most powerful in the world – had outstripped all its competitors to become a worldwide symbol of wealth. This was partly due to a booming economy in America, where international celebrities (notably Hollywood film stars) felt that they had 'made it' only when they could be seen riding around in the latest Cadillac. It coin-

△ ***Nothing glittered quite so much in 1953 as the fantastic Cadillac Eldorado. Owner: Carl Riggins.***

cided with a particularly flamboyant period in American history when the status of cars – and their owners – was frequently rated purely on size and the quantity of chromium plating. In such spheres, nothing glittered quite so much as master stylist Harley Earl's 1953 Eldorado, which cost twice as much as the 'regular' Cadillac series 62 convertible on which it was based.

In those days American car manufacturers changed the appearance of their models every year to encourage customers to buy the latest model regardless of other considerations. Although the golden name Eldorado was transferred to numerous new models, then, Cadillac could never produce quite such an amazing machine again.

This was despite a continuing tradition of technical innovation – Cadillac had produced the world's first synchromesh gearbox in 1929 – in which the Eldorado Brougham limousine received air suspension in 1957 and started a trend (along with Lincoln) for four-headlamp lighting systems.

By 1970, the famous V8's capacity was up to a gargantuan 8.2 litres, but shrank to 7 litres in 1977 in the wake of the 1974 energy crisis, and the engine was eventually made to run on six, or even four, cylinders as a means of economy in 1981! But no matter what the size of the engine – there has even been a 1.8-litre model – Cadillac is still the car to which General Motors expects every American to aspire.

SUMMARY

Plus points

Outrageous styling, symbolic of an age.

Durability.

Surprising performance.

Wallowing comfort if you are not dazzled by chrome interior.

Minus points

Corrosion or trim deterioration.

Non-originality, especially of mechanical components.

CHEVROLET

During the great American boom of the late 1940s, people rushed out to buy anything new, especially cars, after years of limited choice and supply during the Second World War. As a result, car manufacturers competed not so much to make their products more attractive, but to increase production to meet what seemed an insatiable demand. General Motors, in particular, paid little heed to people who complained that its products were boring.

However, when potential customers started paying a premium for European sporting cars like the M.G. TC and Jaguar XK120 (see page 84), General Motors had to sit up and take notice. And so the Chevrolet Corvette was born, in an attempt to counter a possible European sports car invasion.

STING RAY

YEAR MADE: 1966.
ENGINE: V8, 6997 cc, overhead valves, pushrods, 425 bhp at 6400 rpm.
FRAME: Twin girders, cross-braced.
PERFORMANCE: 140–152 mph (225–245 km/h), according to axle ratio.
CURRENT PRICE GUIDELINE: * *
Rare, appreciating model.
PROBLEMS TO LOOK FOR:
Mistreatment, non-originality, chassis corrosion in cold, damp climates.
CLASSIC CAR STATUS: * * Set to rise.

The threat to total sales posed by European sports cars after the Second World War was tiny but, in the tradition established by company founder Alfred P. Sloan, any threat had to be taken seriously. So General Motors responded to the imports on two fronts: it produced the Cadillac Eldorado (see page 43) and, at the same time, introduced its own sports car in 1953, the Chevrolet Corvette. Happily, the Corvette has since continued to prove that Chevrolets do not have to be boring.

The project's chief backers, stylist Harley Earl and chief engineer Ed Cole, had won support from General Motors' normally very conservative senior executives for what was to be the world's first plastic car, on the basis that it would be a mobile testbed for advanced technology. The plans including making a body from glass fibre, which had been used for Second World War patrol craft called Corvettes; hence the car's name. General Motors' hierarchy felt safe in going ahead because it was sure that anything that might go amiss with a 'sporty car' would not affect the normal run of stolid saloons.

The Corvette has continued in its 'testbed' role to this day: only recently it pioneered plastic leaf springs. But there have been some notable models over the years, not the least the Sting Ray series with its early split window and late 427 (cubic inch) power unit which placed these cars amongst the most awe-inspiring creations of the 'muscle car' era.

Budgets for Corvette development were never large, particularly in the early days, because it was not visualized as a high-volume seller. In fact, one of the reasons for using glass-fibre bodywork was that it did not need the heavy investment in tooling associated with mass-produced steel bodies and could be changed quickly as fashion dictated. The rest of the car would have to come from the corporate parts bin.

▽ ***For years the Corvette was America's greatest sport car. This 1967 Sting Ray is owned by Russell Schachter.***

But General Motors soon found that the Corvette needed a distinct character of its own. Early models looked racy, but were underpowered with far from sporting performance, so they did not sell well until 1956 after receiving a modern V8 and manual transmission.

By then development engineer Zora Arkus-Duntov – a keen competition fan with Allard – was lining up a Sebring SS prototype. This was killed, however, by General Motors' rearguard of ageing executives who feared that using the Corvette in professional racing would lead to more deaths on the roads. The Sebring SS was then successfully remodelled as the Sting Ray in 1963 for the less arduous, and presumably safer, amateur production-car racing.

Americans loved the styling – by designer Bill Mitchell – and sales went up by 50 per cent as engines grew ever larger and more powerful, ranging from a mere 327 cu in (5.3 litres) and 300 bhp, to the 427 cu in (7 litres) with 425 bhp and a massive 465 lb/ft of torque in the all-time greatest muscle car of 1966. Looming environmental and safety regulations saw off the Sting Ray in 1967, and Corvettes became much less interesting after that. A longer-wheelbase fourth-generation Sting Ray reached 454 cu in (7.4 litres) by 1970, but lost power due to emission controls after that, bottoming out at 205 bhp by 1975, as the western world suffered its first oil crisis. Also, in around 1970, the Corvette's best friends, Cole, Mitchell and Arkus-Duntov, all retired, ending any hopes of more advanced and sophisticated concepts.

SUMMARY

Plus points

- Dragster-like acceleration.
- Better brakes than earlier models.
- Understressed mechanics.
- Imaginative styling.
- Durable body.

Minus points

- Prehistoric handling.
- Under-tyred by modern standards.

△ ***Eventually even the Corvette had to change. A breakthrough in design in 1983 led to the latest-generation cars.***

CORVETTE

YEARS MADE: From 1984.
ENGINE: V8, 5736 cc, overhead valves, pushrods, 205 bhp at 4200 rpm.
FRAME: Perimeter frame, braced by hoops and bulkhead, together with subframes.
PERFORMANCE: 136 mph (219 km/h).
CURRENT PRICE GUIDELINE: * * *
Highly priced outside America.
PROBLEMS TO LOOK FOR:
Mistreatment.
CLASSIC CAR STATUS: * * Far more sporting than average antiseptic American car.

It was not until 1983 that engineering design perked up and the Corvette entered a fifth generation with its new plastic springs and careful – if not entirely successful – attempts to make it handle, at last, like its European sports car rivals.

Although the new Corvette followed the styling theme of the preceding years, it was clearly influenced by Ferrari's outstandingly attractive 308 GTB and Porsche's 928S (see pages 71 and 145). The Corvette's profile represented a cleaner version than that of the Ferrari, despite being front-engined, rather than mid-engined, whereas, from the front – and at certain angles from the back – it matched the Porsche very closely, although it had a cut-off, Kamm-styled tail. The Porsche and the Corvette had even more in common, notably a similar-sized V8 engine mounted in the front.

Chevrolet could be seen to be moving towards the Ferrari camp, however, and went on to show a prototype mid-engined Corvette Indy featuring a more futuristic development of the same European-inspired lines. This prototype had progressed to full-scale testing by 1986.

SUMMARY

Plus points

Determined attempt to produce a European-style sporting car.

Responsive steering.

Good brakes and grip.

Minus points

Difficult to use transmission.

Power-sapping emission controls compared to Corvettes of old.

CHRYSLER

The British-inspired sports car invasion of the American market in the late 1940s and early 1950s alerted home manufacturers to a demand from young drivers for cars as an entertainment medium as well as for plain reliable transport.

As a result, General Motors produced the Chevrolet Corvette (see page 46) and Ford introduced what it considered was a sporting car, the Thunderbird.

Chrysler, as ever, was lagging third in the great US sales race. While it could not afford to make a completely new car for such a small market, it did not want to be seen to drag its feet on the prestigious matter of performance – some hard thinking was called for by its designers.

CHRYSLER C300

YEAR MADE: 1955.
ENGINE: V8, 5394 cc, overhead valves, pushrods, 300 bhp at 5000 rpm.
FRAME: Twin girder.
PERFORMANCE: 125 mph (201 km/h).
CURRENT PRICE GUIDELINE: * *
Appreciating with age, rare even in America, almost unknown outside.
PROBLEMS TO LOOK FOR: Misuse, old age, non-originality.
CLASSIC CAR STATUS: * * Appreciating as legend spreads wider.

In the face of competition from home and abroad, Chrysler decided to produce a limited-edition saloon car using parts from various other mass-production models that would outperform the Corvette and Thunderbird. Such an inherently heavy car would need a mighty powerful engine but, fortunately, Chrysler had done a lot of research in this area while producing tanks during the Second World War.

In the immediate post-war period the basic American philosophy ran that the bigger a car, the faster it went. In this context, the three main manufacturers began to design bigger and better V8s with overhead-valve configurations replacing the more conservative and less efficient, but cheaper-to-produce side-placed valves. Initially, General Motors led with the 160 bhp overhead-valve Cadillac. But Chrysler had realized that it was possible to go one stage further and build an even more efficient engine, with hemispherical – or 'half-moon' – combustion chambers to take full advantage of overhead valves, providing that high-quality petrol was

▽ *Chrysler's 300B model of 1956 had an even larger hemi-head engine than the C300 on which it was based.*

available. Quite simply, a 'hemi-head' engine could produce at least 10 per cent more power on high-octane fuel than an equivalent engine which had combustion chambers shaped in a cheaper way, requiring fewer working parts to operate the valves.

When it went ahead with development of the hemi-head engine, Chrysler was gambling that oil companies would produce higher-octane fuel to improve their image. Another risk area was strength, because a hemi-head engine needed more working parts, such as a double row of rocker shafts to work the valves, and a massive head casting to contain it all. Just to be on the safe side, it made the rest of the engine as big and strong as the head, and this also raised the cost. The resultant engine was big, heavy and expensive, but it obliged with 180 bhp against the equivalent Cadillac's 160.

Sports car racer Briggs Cunningham even switched from Cadillac V8s to Chrysler's Hemi for his Le Mans cars, which did Chrysler's image no end of good. Its engineers learned a lot from the Cunningham experience, extracting as much as 400 bhp from Hemi engines intended for Indianapolis racing. This frightened the conservative Indianapolis establishment so much that engines like the Hemi were banned. The units, however, rapidly became popular in America's saloon car racing series, called NASCAR, and eventually provided a winner in 1954, despite the fact that the car in question looked square and stodgy and was far from aerodynamic.

Spurred on by the trendy two-seater Corvettes and Thunderbirds which had just been introduced, Chrysler decided to hit back with a stylish new saloon which would offer four seats, go faster and handle better. This meant building a road – or 'street' – 300 bhp version of the Indianapolis Hemi.

Chrysler was lucky in that it had a brilliant stylist, Virgil Exner, who produced an attractive new two-door saloon, having persuaded his management to make do with only a little chrome, so that it had a 'thoroughbred' look. This was what was to become the legendary Chrysler C300 (for Competition 300 bhp) coupé of 1955, which won almost everything in sight.

After that the 300 model switched to a designation with a letter suffix to encourage customers to buy a new one each year. Thus, the 1956 model became, confusingly, the 300-B! This had an even larger hemi-head engine, of 354 cu in (5.7 litres) as against 331 (5.4 litres), producing between 340 bhp and 355, depending on its state of tune. The power of the Hemi was increased even further to 392 cu in (6.4 litres) and 375-390 bhp for the 300-C in 1957 before the Automobile Manufacturers' Association of America persuaded its members – including Chrysler – to stop using speed as a way of selling cars. This was ostensibly on the grounds of safety, but effectively on the grounds of cost. The Chrysler 300, which went on to an L-type in 1965 and a plain 300 in 1966, was never the same again.

SUMMARY

Plus points

Rarity value.

Immense performance.

Durability.

Easy to drive with two-speed automatic gearbox.

Minus points

Prone to eventual corrosion in damp climates.

Original parts can be hard to find.

CISITALIA

Piero Dusio became a brilliant salesman and then a clothing manufacturer when his career as a professional footballer in Turin ended, and by the mid-1930s, he could afford to drive Grand Prix cars as a hobby. He made another fortune supplying the Italian army with its uniforms and the civilian population with bicycles. His operation, which also involved banking, hotels and sports goods, bore the name Consorzio Industriale Sportivo Italia ('consortium of sporting industries'), or Cisitalia for short. By 1944, Dusio realized that the war would soon be over and that he would need something to publicize his empire. What could be better than a racing car, which would also fit in with his hobby?

CISITALIA TYPE 202

YEARS MADE: 1947–9.
ENGINE: In-line four-cylinder, 1098 cc, overhead valves, pushrods, 55 bhp at 5500 rpm.
FRAME: Tubular spaceframe.
PERFORMANCE: 97 mph (155 km/h).
CURRENT PRICE GUIDELINE: *** Very rare, good examples difficult to locate.
PROBLEMS TO LOOK FOR: Misuse, non-originality, check that numerous special parts are in good order.
CLASSIC CAR STATUS: ****
Greatly undervalued due to relative lack of knowledge of its historical significance and to its diminutive size.

Dusio's first step was to hire Fiat's brilliant Dante Giacosa, part-time, as his designer. Initially, the Cisitalia racing car had to be based on readily available Fiat components because of wartime shortages. But there was plenty of high-quality steel tubing available in Dusio's bicycle works, and Giacosa celebrated this abundance by designing the world's first successful 'space frame' for a racing car. This was based on his earlier aeronautical work with Fiat and took its name from the fact that, when viewed as a whole, there was more space in the frame than solid metal. But the spider's web of tubing – bearing a distinct resemblance to the frame of an aircraft's fuselage – which made up the Cisitalia's chassis was not only extremely rigid, but very light as well because of all the space between the tubes.

Dusio was also well-stocked with machine tools (from another sideline, making car-repair equipment), so special engines were constructed around the basic Fiat cylinder block, producing far more power than the original. There was even a sophisticated pre-selector gearbox which enabled the driver to keep both hands on the wheel during hectic cornering, as he could change gear with his foot.

Fiat gave a lot of technical encouragement and in 1945 even allowed Dusio to hire full-time another leading designer from its aircraft section, Giovanni Savonuzzi, to develop the Cisitalia. Dusio responded by winning the first closed-circuit race held after the war, the Coppa Brezzi, with his new car which he had called the D46,

because it was 1946. But a proposed world tour, during which top drivers in a fleet of Cisitalia D46s would introduce motor racing to far-flung countries, collapsed after the first round in Cairo when the event attracted disastrously small crowds.

But meanwhile Giacosa had designed a two-seater sports car based on the D46, so Dusio turned his attention to that. Early attempts by specialist coachbuilders at producing bodies on this chassis were disappointing, until Savonuzzi came up with a brilliant new streamlined coupé with full-width sides, again following aircraft practice. It also featured an aerodynamic spoiler and prominent fins on its rear wings which were to revolutionize the shape of post-war cars, especially those from Cadillac, which turned tail fins into a great selling point. General Motors also quickly adopted the ventilation portholes which appeared on the sides of another Savonuzzi Cisitalia coupé, creating a further styling trend.

By 1947 these beautiful Cisitalia coupés were performing well in road races like the Mille Miglia around Italy, despite having engines of only 1100 cc. Dusio's sights were raised once more and it needed little encouragement from his leading works driver, Piero Taruffi, for him to aim for the top in Grand Prix racing. To his eternal regret, Savonuzzi suggested contacting surviving German engineers who had been responsible for the world's fastest racing cars before the war. Dusio became obsessed with the project and paid his entire post-war profits of 1 million French francs as a ransom to liberate Ferdinand Porsche, who had been jailed in France as a war criminal. The designer of the pre-war Auto Union (and wartime tanks) then initiated an incredibly sophisticated, but still-born, machine which was so far ahead of its time that it had bankrupted Dusio (and led to the resignation of the more practical Savonuzzi) by 1949.

Meanwhile, Dusio's racing team manager, Carlo Abarth (see page 8), concentrated on developing his own business with Rudolf Hruska, as Prof. Porsche's representative in Italy. And Porsche, back at home in Austria with his son, Ferry, started work on a sports car based on the Cisitalia system of using basic production parts to make a high-performance machine (see page 141). In their case, however, they used the readily available German Volkswagen, rather than the Italian Fiat.

△ ***The Cisitalia Type 202 had full-width bodywork and a spaceframe and standard car components that inspired Porsche. Owner: Paul Kunkel.***

SUMMARY

Plus points

- Rarity value.
- Historical value.
- Numerous trend-setting innovations.
- Wonderful performance.
- Excellent handling.
- Beautiful construction.

Minus points

- Non-existent spare parts.
- Expensive construction.
- High restoration costs.

CITROËN

Army officer André Citroën was not only a brilliant engineer, but a gambler, too. So much so that, when he saw peasants cutting wooden gearwheels in a herringbone formation to make them last longer, he left his career in the armed forces and staked everything he could raise on a factory to produce metal cogs of the same pattern. The idea worked so well that, within five years, Citroën's gearwheel factory was one of the biggest in France. To this day, the company he founded works under the symbol of the double chevron.

As a captain in the French army reserve during the First World War, Citroën realized the inadequacies of his country's shells and raised enough government backing to buy 30 acres of land on the Quai de Javel in Paris for a munitions factory. When the war ended, Citroën switched his new factory to car production to make use of the gears he produced there.

By 1922, with the gambler's gift for knowing what was wanted – a simple, solid, easy-to-assemble automobile – and an equally finely honed grasp of the laws of supply, demand, pricing and publicity, Citroën was selling more than 100,000 cars a year. Inspired by American mass-production methods, in 1925 Citroën introduced France's first all-steel-bodied car, the C series, to put his firm on a par with established giants Peugeot and Renault.

TRACTION AVANT

YEARS MADE: 1934–57.
ENGINE: In-line four cylinder, typically 1911 cc, overhead valves, pushrods, 56 bhp at 4000 rpm.
FRAME: Integral.
PERFORMANCE: 70 mph (113 km/h).
CURRENT PRICE GUIDELINE: * Underrated in view of historical importance.
PROBLEMS TO LOOK FOR: Corrosion typical with old age, non-originality, mistreatment, general wear.
CLASSIC CAR STATUS: *** One of the most advanced mass-produced cars ever made.

Sadly, despite his amazing early successes, André Citroën also concentrated much of his energy on gambling away millions of francs at Deauville and Monte Carlo. By the time he realized that he needed a new car to maintain his hold on the market, and had put the revolutionary Traction Avant into production in 1934, it was too late. There were teething troubles that caused a sensational new automatic transmission to be abandoned and the cash ran out. His major creditor, Michelin, bought him out and Citroën died, dispirited, in 1935.

The Traction Avant, however, became his memorial. It combined for the first time front-wheel drive, one-piece (or monocoque) chassis and body construction, torsion bar suspension and an overhead-valve engine with removable cylinder sleeves to make reconditioning easier. It proved to be such a good car that, by the Second World War, 250,000 had been sold.

After the war, the Traction Avant carried on in the same vein, proving a marvellous investment for Michelin and making a mockery of the initial development costs which had helped bankrupt Citroën. It also continued to be a technical pioneer, becoming the first car to use a mass-produced radial-ply tyre – the Michelin X – which gave improved handling and wear characteristics. This remarkable car, immortalized by the fictional French detective Maigret, was produced until 1957.

▽ The Traction Avant bankrupted André Citroën but became a national institution. Owned by Alec Bilney.

SUMMARY

Plus points
- Technological marvel.
- Ultimately usable once in good order.
- Easy to restore if not badly corroded.

Minus points
- Relative scarcity of parts.
- Heavy steering.

2CV

YEARS MADE: From 1948.
ENGINE: Horizontally-opposed two-cylinder, typically 435 cc, overhead valves, pushrods, 24 bhp at 6750 rpm.
FRAME: Steel platform.
PERFORMANCE: 49 mph (79 km/h).
CURRENT PRICE GUIDELINE: * Will remain a bargain for years to come.
PROBLEMS TO LOOK FOR: Corrosion, mechanical and bodily neglect.
CLASSIC CAR STATUS: * Likely to appreciate when model goes out of production.

△ ***Michelin manager Pierre Boulanger created the immortal Citroën 2CV as a basic motor car. Owner: Bob Hughes.***

Under Michelin manager Pierre Boulanger, Citroën had been working on a car aimed specifically at the peasant-farmers who made up a substantial proportion of the French population. Boulanger's precise specification demanded a car with 'four wheels under an umbrella that can, when loaded with eggs, cross a field without breaking one of them'.

The result was a machine that, although very different from the Traction Avant, became equally famous and sold just as well from its introduction in 1948: the 2CV (for *deux chevaux*, or 2 horsepower). Like the Austin Seven, 26 years before (see page 26), it created a new class of budget-conscious motorists, comforted by advanced interconnected suspension that sailed over the worst surfaces imaginable on French country roads and tracks. Much modified, it is still in production today.

SUMMARY

Plus points

Durability.

Economy.

Easily repaired.

Good availability of spares.

Minus points

Very slow.

Tendency towards chassis corrosion.

DS19

YEARS MADE: 1955–68.
ENGINE: In-line four-cylinder, typically 1911 cc, overhead valves, pushrods, 83 bhp at 4500 rpm.
FRAME: Integral.
PERFORMANCE: 95 mph (153 km/h).
CURRENT PRICE GUIDELINE: * Will appreciate eventually.
PROBLEMS TO LOOK FOR: Extensive corrosion, deterioration of hydraulic system.
CLASSIC CAR STATUS: *** Set to appreciate.

It was the tremendous advance in aeronautics technology between 1939 and 1945 that helped make Boulanger's second post-war car a reality. This large, volume-produced saloon, intended to replace the Traction Avant, was to include an engine-pumped, high-pressure hydraulic system which would operate the suspension as well as level the ride, provide semi-automatic transmission at last, and power the steering and brakes. Bodywork which was sleeker and more aerodynamic than many contemporary aeroplanes was designed by Faliminio Bertoni, but plans for a new flat six-cylinder engine had to be abandoned because of soaring costs. Consequently, the revolutionary DS19 was launched with the Traction Avant's engine.

More than any other, this car was years ahead of its time. It eventually replaced the Traction Avant in 1957, when the company was able to launch a cheaper and less sophisticated version, the ID, at the same time. A more luxurious DS – the Prestige – even had a radio telephone in 1958! Estate-car and then ambulance versions followed to provide even more variants than the Traction Avant. Continual development extracted enough power from the old engine without affecting its economy so that, with a new series of similar engines, the DS was able to sustain its reputation for speed and comfort well into the 1970s.

SUMMARY

Plus points

Becoming rarer.

Superb advanced technology for age.

Perfectly practical for everyday use once in good condition.

Easier to maintain than it looks.

Minus points

Corrosion.

Cost of parts—particularly in hydraulic system—can be more than value of car.

▽ ***Citroën's DS19 – superb streamlining, hydraulic suspension and semi-automatic transmission.***

SM

YEARS MADE: 1970–5.
ENGINE: V6, typically 2670 cc, two overhead camshafts, 170 bhp at 5500 rpm.
FRAME: Integral.
PERFORMANCE: 137 mph (220 km/h).
CURRENT PRICE GUIDELINE: * *
Vastly underrated.
PROBLEMS TO LOOK FOR: Corrosion, deterioration of hydraulic system, camshaft chain tensioner problems on original-specification engines.
CLASSIC CAR STATUS: * * * Likely to improve.

Citroën had always been conscious of the need for a power unit as exotic as the rest of the DS19, but could not afford the investment until it managed to acquire a controlling interest in the famous Italian high-performance car maker, Maserati.

The result was everything that Citroën had dreamed of – the SM, claimed at the time to be the most technically advanced car in the world. The ultra-streamlined body, hydropneumatic suspension, five-speed gearbox and front-wheel drive were typically Citroën, while the fuel-injected V6 power unit – based on a race-bred V8 – was typically Maserati.

In its time, the Citroën SM was the ultimate luxury car, but its life was short as western economies reeled under the shock of the world's first energy crisis in the winter of 1973, rendering many exotic cars on the market almost unsaleable.

SUMMARY

Plus points
Outstanding comfort and performance.
Eventual rarity value.

Minus points
Corrosion.
Cost of repairs, particularly to hydraulics and engine if not converted from original specification.
Camshaft chain tensioner.

CX

YEARS MADE: From 1974.
ENGINE: In-line four-cylinder, typically 1995 cc, single overhead camshaft, 106 bhp at 5500 rpm.
FRAME: Integral.
PERFORMANCE: 109 mph (176 km/h).
CURRENT PRICE GUIDELINE: * *
Excellent value.
PROBLEMS TO LOOK FOR: Corrosion in older models, deterioration of hydraulic system.
CLASSIC CAR STATUS: * Will improve eventually.

In 1971 Citroën introduced an advanced new DS-style small car, the GS, but this did not prevent the company being taken over by Peugeot in the fight for economic survival in 1975. By then the DS was about to be replaced by an even more advanced development, the CX2000, which used much of the technology developed in the SM, but in what amounted to a modernized DS bodyshell. This car became available in as many variants as its predecessor and is still one of the fastest, most economical and luxurious in the world. Citroën has gone on to launch a medium-sized BX, which makes use of ultra-modern lightweight materials in its construction, and a small AX; all retain the individuality that has always been as much a hallmark of Citroën as the sign of the double chevron.

SUMMARY

Plus points
Character.
Ride.
Appearance.
Steering.
Comfort.
Value.

Minus points
Poor ventilation.
Corrosion.
Hydraulic problems in elderly examples.

◁ ***The Citroën SM supercar was produced by an alliance with Maserati. Owner: The Patrick Collection.***

◁ ***Much of the technology developed in the Citroën SM was transferred to the CX saloon and estate cars.***

CORD

Errett Lobban Cord was a classic salesman, perhaps the greatest America has ever known.

He was also a financial miracle-worker, going bankrupt three times before bouncing back to make fortunes outside the motor industry in land, an early pop music radio station and, finally, uranium.

But his story could have been very different, for he nearly struck gold with his first motor car – the L-29.

MODEL 812

YEARS MADE: 1935–8.
ENGINE: V8, 4730 cc, side valves, 125 bhp at 3000 rpm.
FRAME: Integral.
PERFORMANCE: 100 mph (161 km/h).
CURRENT PRICE GUIDELINE: * * * Very rare, could appreciate rapidly just because it is a Cord.
PROBLEMS TO LOOK FOR: Corrosion, old age, mistreatment.
CLASSIC CAR STATUS: * * * Styling should lift it close to the status of the later coffin-nosed models.

The L-29 was so called because it made its début in 1929 with power from Lycoming, the aero-engine wing of the Auburn car manufacturing company, which Cord owned (see page 24). He had acquired the decaying Auburn company of Indiana and the legendary Duesenberg high-performance car makers, as part of a financial package put together for backers after successfully running an agency for luxury cars made by the American firm, Moon.

Cord knew little about cars other than how to sell them. His engineering ideals were inspired far more by what was fashionable than by what was practical. From Auburn's base in Indiana, it was easy to see that front-wheel drive had a future as Miller racing cars were toppling conventional rear-wheel-drive Duesenbergs at Indianapolis. Cord knew that he could sell the advanced front-wheel-drive concept to the American public if it could be housed in a car which looked different. Dispensing with the conventional propeller shaft between the transmission and rear axle would allow the seats to be mounted lower and the car to be made to look more rakish and aggressive. Costs could be saved, wherever possible, by using components made by Cord's companies so that more money could be spent on glamorizing the car.

So this first Cord had a cheap Lycoming straight eight-cylinder engine that was shared with the cost-cutting Auburn made in the same works. It also pioneered the cruciform bracing that was to become a hallmark of American cars for the next 25 years and which was the result of desperate last-minute efforts by engineers to stop the long chassis rails from whipping. The car's performance was indifferent, due to the fact that the L-29 weighed more than 2 tons.

It handled well, however, and displayed amazing cornering ability because of its front-wheel drive. The L-29 was not America's first road car with that type of driveline – a Ruxton beat it to that distinction by a month – but it was not for want of trying by Cord. He forced through development of the L-29 at a tremendous rate, partly by terrifying his staff and partly by refusing to take no for an answer.

And when it came out, the motoring press was wildly excited by this newcomer to a class dominated by stolidly conventional Packards, Chryslers and Cadillacs. The Cord's glamorous body and front-wheel drive more than compensated for the mundane mechanical components. Top coachbuilders were enchanted by the opportunity to mount long, low, rakish bodies on a chassis only 16 in (406 mm) from the ground at a time when the average was half as high again.

Unfortunately for Cord and the town of Auburn, Indiana, the public was not so impressed, chiefly because the L-29 was priced at upwards of $3000, with which you could buy two luxurious houses at that time. People with that sort of money wanted a car which was perfect in every way. Even slashing $800 off the price failed to boost sales, which had been hard hit with the collapse of Wall Street two months after the L-29's launch. From prophesying a demand for at least 10,000 a year, Cord had to be content with 4,500 over three years.

Cord hit back with the 810, designed by legendary engineer Augie Duesenberg, with a new low-line body by

Duesenberg's distinguished stylist, Gordon Buehrig. The Lycoming engine still produced the same power, but the total weight was cut by a quarter and the frame and body were built together to make them more rigid. It also had independent front suspension, and, once more, it was the styling which was sensational. The headlights could be cranked back by handles on the dashboard to retract into the silhouette of the front wings, which surrounded one of the most distinctive noses ever seen on a car. The radiator grille had eight horizontal bars running back along the bonnet sides to the Mercedes-style exposed exhaust pipes. With the grille surmounted by a massive prow, the effect was stunning!

Costs were saved in an ingenious way on a saloon version by making the doors on each side with identical profiles to their opposite numbers – an early example of techniques which would become common in the motor industry.

The new Cord, in 810 and 812 forms, represented a tremendous advance over the L-29, but it could not overcome the resistance of a public who had never seen anything like it before, so it sold only spasmodically in the years before the Second World War.

▽ Cord's famous coffin-nosed 812 became one of the most sought-after classics. Owner: Ray Hunter.

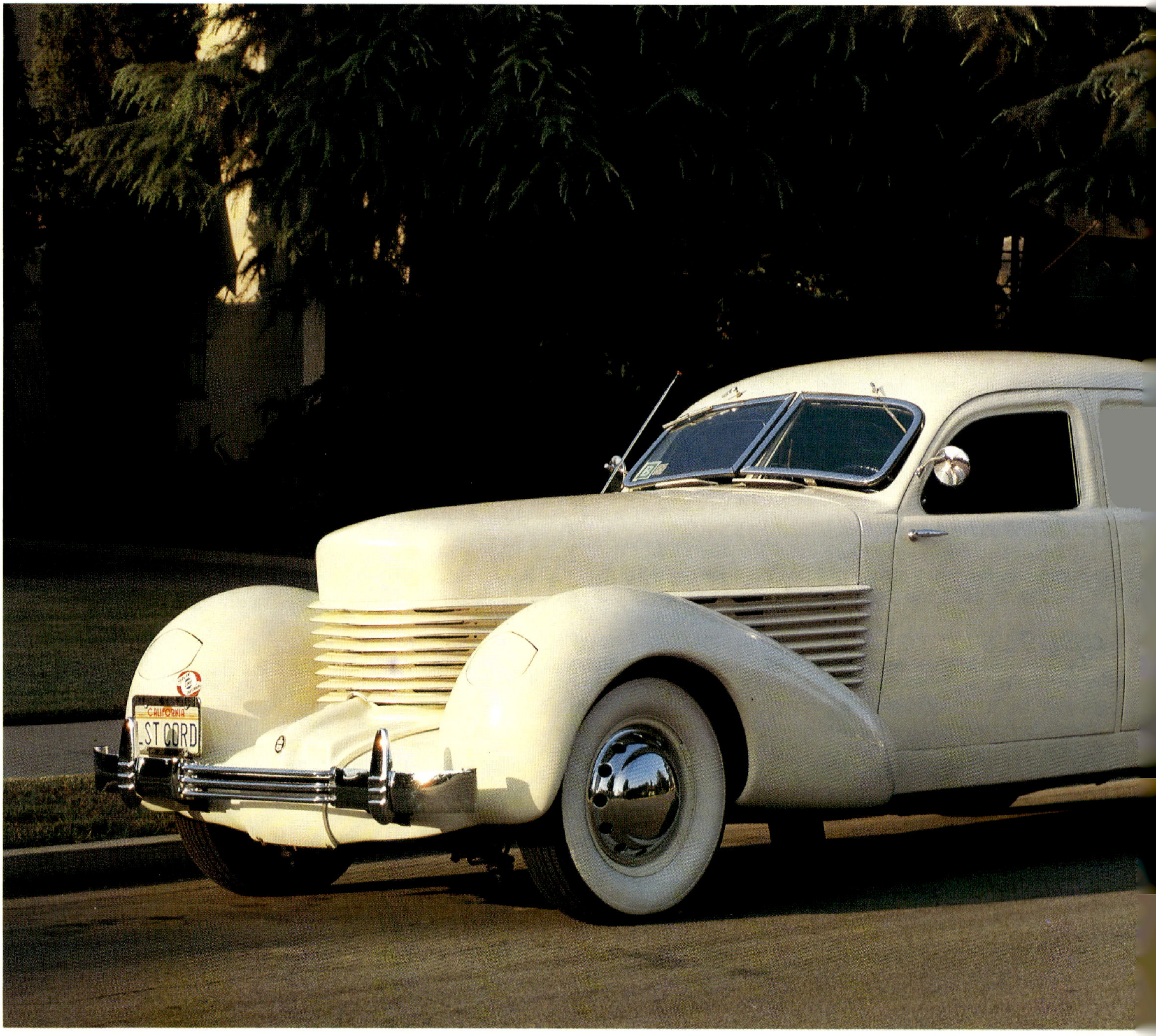

SUMMARY

Plus points

Fantastic appearance.

Technical novelty.

Massive pre-war construction.

Understressed mechanical components.

Minus points

Lack of spares.

Heavy to drive.

High restoration costs.

△ ***The modern Daimler Double-Six, of which this is a coupé version, took its name from the pre-war classic.***

DAIMLER

The names Daimler and Benz are the oldest in the motoring world. Karl Benz is credited with producing the first motorized carriage, a three-wheeler, in 1885. At about the same time, Gottlieb Daimler was founding the company which would be Benz's greatest rival, at Bad Cannstatt, only some 60 miles (97 km) away in Germany. Daimler and Benz never met, but their products were among the most advanced in the new motor industry, and the Daimler Motor Syndicate was formed in England in 1893 to exploit Gottlieb Daimler's patents. However, the English and German concerns were soon following totally different paths.

DOUBLE-SIX

YEARS MADE: 1972–84.
ENGINE: V12, 5342 cc, two overhead camshafts, 295 bhp at 5500 rpm.
FRAME: Integral.
PERFORMANCE: 145 mph (233 km/h).
CURRENT PRICE GUIDELINE: * to *** depending on age. Undervalued if in good condition, set to appreciate rapidly when similar models are finally out of production.
PROBLEMS TO LOOK FOR: Corrosion may be found on old and poorly maintained models.
CLASSIC CAR STATUS: *** Set to appreciate dramatically.

One of the first customers of the Daimler Motor Syndicate was King Edward VII, then Prince of Wales, who forged links between Daimler and the British royal family which were to last for more than 50 years. The company soon became famous for its large, heavy and luxurious cars, which were ideally suited to state occasions. In this context, sleeve valves and underslung worm-drive rear axles were to be standard equipment for many years because they were so strong and silent and combined well with cars which had, typically, 12-cylinder engines of as much as 7.1 litres capacity. The British Daimlers also pioneered the fluid flywheel transmission with steering column selector which would remain standard from 1932 until 1956.

Oddly, there was one Daimler in this period which could be described as sporting. It featured a special Park Ward coupé body on an underslung version of the fabulous Double Six chassis, which normally saw service hauling massive limousine bodywork. With a radiator cap only 3ft 6in (1067 mm) from the ground and a 7.1-litre V12 engine tuned to give more than 150 bhp, this 1930 car had an appearance at least as stunning as any of Errett Lobban Cord's contemporary creations! Its price was every bit as fantastic as its looks, however, and very few were sold.

After the Second World War, the British Daimler company carried on with limousines and similar luxury machines, by now using more efficient and conventional valve gear, until it attempted to boost sales by introducing a much more sporting Conquest saloon in 1953. Then a high-compression engine was fitted for a model known as the Conquest Century because it could exceed 100 mph (161 km/h), and there was also a sports two-seater.

Sales of these high-quality cars with their traditional lines still flagged, however, in the face of opposition from rivals like Jaguar, which offered a more advanced engine, inspired styling and a lower price tag. Jaguar's recent reputation had been founded on the success of its sports cars, so Daimler hired Edward Turner, who had enjoyed great success with Triumph motor cycles, to design a new range of power units.

The first of these engines, a lovely little 2.5-litre V8 based on four Triumph twin-cylinder motor-cycle engines, made its début in a 120 mph (193 km/h) sports car, the SP250, in 1959. This was an amazing machine in that the nose bore a distinct resemblance to that of the trend-setting Citroën DS (see page 51), while the fins at the back were distinctly transatlantic. The body was also quite unlike any Daimler that had gone before it, in that it was made, Chevrolet Corvette-style, from glass fibre (see page 44). The faithful could hardly believe that they were being offered a plastic Daimler!

Although the price was competitive with that of a Jaguar, the SP250 was such a complete departure from Daimler's established practice that sales suffered as a result. A formidable 4.5-litre version of the SP250's engine was then fitted to Daimler's big (steel-bodied) Majestic saloon to give it a performance almost as good as the best Jaguar – so in 1960 Jaguar simply bought Daimler out!

These Daimlers continued in production for a while until the SP250 – which had been no match for Jaguar's E-type (see page 89) – disappeared so that its engine could be used in a Daimler version of the best-selling Jaguar Mark II saloon (see page 88). Then after 1967, when Jaguar became part of the British Motor Corporation, ranges had to be rationalized and Daimlers became simply Jaguars with slightly more up-market trim.

There was a happy ending for the marque, however. When Jaguar introduced first the XJ6 saloon (see page 89) and then the XJ12 with a wonderful new 12-cylinder engine, the top-of-the-range Daimler – also called the Double Six – was proclaimed Best Car in the World by *CAR* magazine, taking a title long held by Rolls-Royce!

SUMMARY

Plus points

Wonderful performance, handling, ride, comfort, silence, security, styling.

Excellent availability of spares.

Minus points

Extensive corrosion on old and neglected cars.

High fuel consumption on earlier models.

DATSUN

M.G. and Jaguar, closely followed by Austin-Healey and Triumph, set the style for post-war sports cars in the great American market while other manufacturers concentrated on mass-production saloons. The Japanese motor industry was almost unknown in the 1950s, Britain's golden era. Many of the Japanese products were simply foreign makes produced under licence, with a total of just 22 cars exported between 1946 and 1955. How the situation was to change!

240Z

YEARS MADE: 1969–73.
ENGINE: In-line six-cylinder, 2393 cc, overhead camshaft, 151 bhp at 5600 rpm.
FRAME: Integral.
PERFORMANCE: 125 mph (201 km/h).
CURRENT PRICE GUIDELINE: * to **
Relatively scarce, for a mass-produced car, outside the United States.
PROBLEMS TO LOOK FOR: Extensive corrosion, mistreatment, mechanical wear, especially in transmission.
CLASSIC CAR STATUS: * Will take some time to appreciate.

The Nissan Motor Company, which was later to export to North America and Europe by the million, concentrated in the 1950s mostly on making its own versions of British Motor Corporation cars. As sales of the 'big four' British sports cars boomed, nobody except the Japanese seemed to want Nissan's Datsun S211, a glass-fibre-bodied open four-seater, remarkably like an Austin-Healey (see page 28).

But when M.G. produced its MGB in 1962 (see page 30), Nissan started to look far more threatening. It followed up instantly with a very similar car, the more sporting two-seater Datsun Fairlady SP310. It was not long before 3500 a year were being exported to America, and Nissan was dreaming of fitting a larger and more powerful engine to make the Fairlady a cut-price competitor for the Jaguar E-type (see page 89). It had to shelve the project for a while, however, when the planned new engine from Yamaha ran into severe problems.

△ ***Datsun's 240Z, inspired by numerous other machines, became the world's top-selling sports car. Owner: J. Gordon.***

But it needed only Nissan's deadly rivals, Toyota, to reveal similar plans, for the new Fairlady project to be re-activated with a vengeance. Happily for Nissan, it had by this time developed quite a powerful 2.4-litre straight six-cylinder engine (bearing a distinct resemblance to a single-overhead-camshaft Mercedes design) for its Datsun Bluebird 510 saloon, which was being produced in large quantities. Along with a four- or five-speed gearbox from the same source, Nissan now had a cheap new power train that could keep costs down. Inexpensive and simple suspension, based on that pioneered by the British firm Lotus, completed the mechanical picture.

Nissan was then able to afford an entirely new bodyshell because it had saved a great deal of money by not having to develop new mechanical parts for the sports car. This body took the form of a fixed-head coupé with a nose like a Chevrolet Corvette Sting Ray (see page 44) and a cabin like that of Jaguar's E-type coupé. For the first time, also, Nissan was producing a car big enough to accommodate broad-beamed Californians, who were, on average, 20 per cent larger than the Japanese, although the car itself occupied no more space than a Porsche 911 – another of Nissan's criteria. The reason for choosing a fixed-head coupé rather than the popular roadster shape, was that American laws were expected to outlaw open cars before long on safety grounds. It is ironic to note that this type of car was exactly what M.G. was trying to sell at the time to replace the outdated Austin-Healey 3000 (see page 129). M.G. failed with the MGC, however, because the Austin engine it was forced to use was so heavy that it ruined the handling.

The result, when the Datsun 240Z, as the new car was called, was introduced late in 1969, exceeded even the wildest dreams of Nissan. For the first time since Jaguar had launched the E-type in 1961, Americans were standing in line for a new sports car. This was because the 240Z was as fast as a Big Healey (see page 30) – which had been killed by new environmental exhaust and safety laws – and cost only half the price of the surviving Jaguar E-type. It was estimated that there were ten people wanting to buy a Datsun 240Z for every car available.

Nissan was thrilled because the success rubbed off on its other models, with many potential customers for the 240Z buying small saloons like the Datsun Laurel as alternative family cars. By the end of the year, Nissan was the seventh largest motor manufacturer in the world, with exports of 300,000 cars a year, chiefly to the lucrative American market.

Within two or three years, however, the 240Z had to change. The same emission laws which had killed off the Big Healey and caused so many problems with the MGC, forced Nissan to fit a larger, less efficient, 2.6-litre engine. This put up the price of the new 260Z sufficiently to make it essential that Nissan incorporate as standard two tiny rear seats so that the car appealed to more affluent people who would not be satisfied with just two front seats. After that the 'Z car', still known as a Fairlady in Japan, became increasingly bigger and heavier as larger and less efficient engines were needed to satisfy the ever tougher emission laws in its biggest market, America. That left the original 240Z as the classic, although 'Z cars' of all types have now sold more than a million, making them the world's top-selling sports cars.

SUMMARY

Plus points

Extremely durable if well treated.

High, usable performance.

No particular spares problem, especially in America.

Minus points

Corrosion in pressed steel body.

Transmission can give problems.

DE TOMASO

Argentinian former racing driver Alejandro de Tomaso, who settled permanently in Italy after falling out with dictator Juan Perón, has become one of the most successful entrepreneurs in the history of motor sport and the specialist manufacture of classic cars. After marrying a fellow racing driver, American heiress Elizabeth Haskell, former OSCA works driver de Tomaso set up in business making prototype cars aimed at capturing established markets, notably that of next-door neighbour Ferrari, the traditional rival of the Maserati brothers who made OSCAs.

MANGUSTA

YEARS MADE: 1966–71.
ENGINE: V8, 4735 cc, overhead valves, pushrods, 300 bhp at 5750 rpm.
FRAME: Fabricated steel backbone.
PERFORMANCE: 155 mph (249 km/h).
CURRENT PRICE GUIDELINE: ** to *** Very rare, good examples difficult to locate. A great investment, providing it is not necessary to drive it.
PROBLEMS TO LOOK FOR: Corrosion, accident damage, transmission wear, electrical deterioration.
CLASSIC CAR STATUS: **** Prices held down by difficulties in using machine.

Once in business, de Tomaso dreamed up increasingly exotic cars, using the numerous specialists who had established themselves around the Ferrari factory to turn his ideas into reality. The one thing such cars lacked was development, because de Tomaso always hoped to sell his designs to a major manufacturer, so that he would have finance to build others himself. By 1965 he had a small sports car – named the Vallelunga after the race track near Rome – heading for small-scale production and boasting a Ford Cortina engine and a backbone chassis like the Lotus Elan (see page 110).

And then a new idea gelled in his fertile brain: why not turn it into his own version of the Ford GT40, which was already being hailed as an instant classic (see page 75). All he would need to do was scale up the chassis, fit the Ford's AC Cobra engine (see page 9), and hire somebody to design a stunning new body. The engines would be no problem: Carroll Shelby, who was running the Cobras in GT racing, was an old friend and could arrange supplies. De Tomaso could not help noticing, however, that, despite Ford's advanced new iron-casting technology, its engine was still very heavy, and that Colin Chapman of Lotus was winning at Indianapolis with alloy-blocked experimental versions. There was not much hope of liberating those engines from Ford, so de Tomaso did a deal with Chapman, and between them they financed their own alloy castings, de Tomaso proclaiming the result as his own brilliant new 500 bhp engine – although it contained no pistons, crankshaft or connecting rods! As ever, he hoped somebody would pay to put it into production.

Meanwhile, the Italian coachbuilder Ghia – which was producing Vallelunga bodies among others – had fallen upon hard times, as its effective owner, Rafael Trujillo, son of the former Dominican dictator, was in jail. With the help of the Haskell family, who ran the giant American electronics firm Rowan Controllers, de Tomaso took over Ghia so that he now had his own organization for designing prototypes.

One of his first acts was to hire the brilliant young designer, Giorgetto Giugiaro, who was seeking a more prominent platform for his ability than the relative anonymity of the old-established Bertone empire. As Giugiaro set to work on the new project – and numerous others – de Tomaso came up with an inspired name: Mangusta, which is Italian for mongoose . . . the small creature which eats Cobras for breakfast.

Many aspects of the Mangusta's design were brilliant, whilst others were

far from practical. But the body was something ultra-special, representing Giugiaro at his best. It had a purity of line that highlighted the mechanical parts which gave the car its power. Of such cars legends are made. Indeed, the Mangusta more than lived up to its name, threatening to consume its occupants as the chassis flexed, leading to wild and vicious handling problems.

But it looked fantastic and could not fail to catch the eye of Henry Ford. Scorned in a bid to take over Ferrari, his company was still suffering from a massive inferiority complex when the Mangusta appeared in 1966. He remembered how the first Henry Ford had failed to buy the Italian exotic car maker, Isotta Fraschini, in 1930. Having an Italian-born wife, Cristina, he was determined to achieve what his grandfather could not, if only to show that Americans did have good taste, and to create new attractions in Ford showrooms at the same time. He needed only to fail again, as Fiat bought Lancia, to swoop on the American-owned De Tomaso, much to Alejandro's delight.

The Mangusta gave Ford problems enough, as did the Pantera which followed, despite a conventional bodyshell to replace the flexing backbone chassis. But Ford was eventually content, having sold off everything except Ghia, which has been retained as a European styling studio of exceptional quality. De Tomaso then took back his more fully developed Pantera and, with government aid, built an empire which embraced his first love, Maserati.

△ ***The Dodge Charger, one of America's greatest 'muscle cars'.***

◁ ***De Tomaso's Mangusta had stunning styling by Giorgetto Giugiaro, although it was far from practical.***

SUMMARY

Plus points
- Fantastic, historic styling.
- Strong engine and gearbox.

Minus points
- Appalling handling can be downright dangerous.
- Body and chassis prone to extensive corrosion.
- Poor build quality.
- Difficult clutch.
- Minimal ground clearance.
- Eccentric electrics.

DODGE

The Great Depression that followed the Wall Street crash of 1929 left the average American man with little money, a poor job if any at all, and survival as his chief ambition. Automobiles figured only in his dreams, and then fleetingly. But as the Depression lifted in the late 1930s, he was once more able to think about buying a new car. The Second World War put paid to such thoughts again for a while, followed by a few years, in the late 1940s, when the floodgates opened and he fought to buy any new car at all. Manufacturers vied with each other and themselves to produce cars in ever increasing numbers, setting the scene for the emergence, in the 1950s, of a new breed of Detroit iron, now called 'muscle cars'. The Dodge Charger was to be one of the greatest of all.

CHARGER

YEARS MADE: 1967–9.
ENGINE: V8, 6981 cc, overhead valves, pushrods, 425 bhp at 5000 rpm.
FRAME: Integral.
PERFORMANCE: 142 mph (229 km/h).
CURRENT PRICE GUIDELINE: * * * Hemi-engined Charger is a rare model, even in the United States; likely to hold value.
PROBLEMS TO LOOK FOR: Corrosion, check that components are original.
CLASSIC CAR STATUS: * * Likely to improve in value with age.

By the early 1950s, the American automobile industry was well set in its ways, to such an extent that big cars had big engines for a big performance, medium cars had medium-sized engines, and cheap cars were just plain cheap. As Chevrolet, Ford and Plymouth concentrated on supplying inexpensive, dependable and docile vehicles for those seeking economical transport, so Cadillac, Lincoln, Chrysler and Packard continued to cater for the wealthy, with Pontiac, Oldsmobile, Buick, Mercury, De Soto and Dodge somewhere in between.

By 1955, when Chrysler had swung into top gear with its Hemi engine (see page 47), and Chevrolet had produced an overhead-valve V8, horsepower had become a highly marketable item. For the first time, cheaper cars could keep up with the top models. Then, over the next ten years, the lower-priced cars forged ahead as a new younger generation entered the car-buying classes which had been dominated before by their far more conservative parents.

The Pontiac division of General Motors was the first to capitalize on this trend by fitting a large-capacity V8 from its Catalina/Bonneville range into the much smaller and lighter Tempest bodyshell. To complete the illusion, it borrowed the name of Ferrari's world championship-winning sports car: GTO (see page 68). The result was incredible. The new Pontiac GTO galloped away from the traffic lights at a pace normally seen only on the drag-racing strips which had become highly popular in America. And because the model was smaller and more sparsely equipped than the Catalina, to keep down the weight, its cost could be kept down, too. Sales soared in what General Motors identified as the 'Youth Market' – customers aged between 18 and 34. In retrospect, it can be seen that the Pontiac GTO was the first of the 'muscle cars'.

Soon all the rivals had adopted this formula, in such notable models as the Chevrolet Impala SS 409, the Buick Skylark GS 400, the Oldsmobile 4-4-2 and the Mercury 202 Sedan 427. Chrysler, which, despite the 1957 industry agreement attempting to ban competition programmes, had been battling mightily with Ford on racetrack and dragstrip, replied with the most fearsome of all such muscle cars. This was the Dodge Charger that had only one object: 'to bite deep into the Pontiac GTO belt'.

The Charger was seen as a way of 'letting the potential buyer recognize the Chrysler-Dodge speed image'. In other words, if the potential buyer could not afford one (or, more likely, could not afford to insure one, as the insurance companies had now caught up with the performance), he could, hopefully, be deflected towards a more mundane Chrysler product. Such models would then be festooned with 'go-faster' goodies. Ideally, these extras would not boost the performance unduly, and thus not affect the insurance, but they would make the car look like a racer.

In this context, the 1967–8 charger acquired notchback 'sport-oriented' styling based around a recessed rear window, with a spoiler formed in the bootlid to keep the back end on the ground. It also had retractable headlights to look like the taped-over items fitted to competition cars. These styling features were then adopted for other models in the next year's range of Chryslers and Plymouths.

The Charger, meanwhile, was offered with a choice of engines varying between 5211 cc, and 7210 cc, all capable of sending it away from a traffic light in clouds of tyre smoke, hitting 60 mph (97 km/h) in as little as 5 seconds. Turning corners, as the insurance companies had realized, could be a trickier proposition, although there were few sharp bends on the average American highway.

The most sought-after Charger had the legendary 426 cu in (7-litre) Hemi, although there was another model with a conventional cylinder head 440 cu in (7.2-litre) V8 which offered even more torque.

Inevitably, the reign of the muscle cars was short. A combination of punitive insurance rates for young customers, and the Vietnam War which took many of them away from America, plus a new class of lighter Pony Cars, made these models extinct by the early 1970s. But as a fast-accelerating marketing exercise, the Dodge Charger was the greatest, being recognized as a classic in the popular 'Dukes of Hazzard' TV series.

SUMMARY

Plus points

Massive straight-line performance.

Reliability.

Durability.

Easy to drive in a straight line on a smooth surface.

Minimal maintenance needed.

Minus points

Early muscle car handling.

Corrosion in areas where roads are salted.

Ultimate-specification examples need high-octane fuel.

DUESENBERG

Immigrant German engineers Fred and August Duesenberg began their business life in America building bicycles, but rapidly expanded into racing cars from their headquarters

near the Indianapolis Speedway. By the First World War they had achieved such a reputation for excellent workmanship and advanced design that they were able to produce Bugatti aero-engines under licence. After that they expanded into road cars and their first proved to be the best in America.

DUESENBERG MODEL A

YEARS MADE: 1920–6.
ENGINE: In-line eight-cylinder, 4188 cc, single overhead camshaft, 100 bhp at 4000 rpm.
FRAME: Twin girder, cross-braced.
PERFORMANCE: 80 mph (129 km/h).
CURRENT PRICE GUIDELINE: * * *
Very rare, even in United States; set to appreciate.
PROBLEMS TO LOOK FOR: Non-originality, old-age deterioration, especially in bodywork.
CLASSIC CAR STATUS: * * * * First of the great American straight-eights.

The Duesenberg brothers' first road-going success was the Model A of 1920, which featured the first straight eight-cylinder engine to be seen in an American production car, and was fast enough – when clad in a lightweight body and fitted with a competition engine – to win the French Grand Prix of 1921 at a record speed of 78 mph (126 km/h). Apart from its distinction of being a straight-eight, the engine was far ahead of any other in the United States at the time, using a single overhead camshaft to produce 100 bhp or more, according to tune. What is more, the chassis was the equal of the engine, complete with America's first hydraulic four-wheel brakes.

By 1926, however, the Duesenberg Motor Company had been taken over by high-flying salesman Errett Lobban Cord, who wanted to produce the finest and fastest car in the world. Meanwhile the Model A appeared in modified form as the Model X with a longer wheelbase, so that Cord could have even sleeker coachwork and cheaper and quieter timing chains to replace the noisy competition-style gear-driven valves, but only a few were made.

SUMMARY

Plus points
Rarity value.
Historic high standard of innovation and engineering.
Wonderfully well made.
Durability.
Ease of driving.

Minus points
Not so beautiful as later Duesenbergs.
High price of restoration.

▽ ***America's finest . . . the Duesenberg Model A, with the first straight eight-cylinder engine and hydraulic brakes to be seen in the US, and a chassis which gave it superb handling.***

SSJ ROADSTER

YEAR MADE: 1935.
ENGINE: In-line eight-cylinder, 6882 cc, supercharged, twin overhead camshafts, 400 bhp at 4750 rpm.
PERFORMANCE: 140 mph (225 km/h).
CURRENT PRICE GUIDELINE: *****+
One of the rarest and most desirable cars in the world, especially for patriotic Americans with access to massive sums of money.
PROBLEMS TO LOOK FOR: Non-originality.
CLASSIC CAR STATUS: ***** The ultimate classic.

There was no compromise with the car which followed the Model X, the Model J of 1928, which came remarkably close to the perfection demanded by Cord. It was bigger, faster, more elaborate and more expensive than any other car in America, and pointed the way towards the Model SSJ, another example of a high-performance car with a price to match!

The Model J's 7-litre eight-cylinder engine had twin overhead camshafts operating four valves per cylinder to give 265 bhp – more than double that of any rival. The brakes were brilliant, of course, using a vacuum servo for assistance, and the steering was among the best ever produced in America, needing only 3.5 turns from lock to lock, despite an overall weight of around 5000 lb. It was also the only American car of its period to be fitted with a rev counter as standard in an instrument panel almost as complete as that of an aeroplane. The Model J even boasted an altimeter!

Its massive twin-girder chassis, braced by numerous tubular cross-members, could be ordered with either a short wheelbase for competition or sporting appeal, or in long-wheelbase form, which delighted all the best coachbuilders. Bodies built by American firms such as Murphy, Bohman & Schwartz, Judkins and Willoughby, Derham, Rollston, Brunn, Weymann, Brewster, Woods, Kirchoff and Walton and Le Baron were some of the most elegant and extravagant seen on any car. But even then, the European equivalents – Barker, Gurney Nutting, D'Iteren, Vanden Plas, Hibberd & Darrin, Fernandez, Kellner, Franay, Figoni et Falaschi, Letourner et Marchand, Saoutchik, Cattaneo, Graber and Castagna – often surpassed them.

The performance of the Model J was just as extraordinary: this enormous car could reach 60 mph (97 km/h) from rest in 8.6 seconds on the way to 95 mph (153 km/h) in second gear and 115 mph (185 km/h) in top. The cost of a complete Duesenberg was also in the same league: as much as $18,000, at a

▽ Duesenberg's greatest . . . film stars Gary Cooper and Clark Gable paid fortunes for the SSJ roadster.

time when men were being paid in cents. In fact, it was one of the most expensive cars in the world.

Despite the Wall Street crash in 1929, Duesenbergs became increasingly opulent, with a supercharged model being added in 1932. This normally used the shorter chassis and was then known as the SJ. Its engine was rated at no less than 325 bhp, achieved at what were very high revs for the day: 4750 rpm. The SJ bested the Model J's 0–100 mph (0–161 km/h) figure of 20 seconds by 3 seconds, and could accomplish that feat in second gear alone!

Three years later, film star Gary Cooper marched into a Hollywood showroom and ordered something special . . . even for a Duesenberg. He wanted the fastest one in the world, and it came with an even shorter chassis, a high-compression engine producing 400 bhp, and roadster bodywork by LaGrande, for a sum equal to his entire earnings from a starring role in *Lives of a Bengal Lancer*. This was the first Model SSJ.

No sooner had Cooper's car been delivered in 1935 than arch-rival Clarke Gable wanted an SSJ – and bought one for a sum not far removed from his fee for the blockbuster, *Mutiny on the Bounty*.

Despite the SSJ's high price, the materials used in the car cost so much that it was effectively a loss-leader to publicize the more mundane products of Cord's empire. And when Cord's empire crashed in 1937, Duesenberg went with it. Nobody could afford to subsidize such extravagance and infrequent attempts since to revive the marque have failed, leaving the originals with the reputation of being the finest cars in the world.

SUMMARY

Plus points
Rarity value.
Fantastic performance.
Amazing engineering for the period.
Wonderfully well made.
Durability.
Ease of driving.

Minus points
Possible high cost of restoration.

△ ***The Facel Vega HK500 dream car was created by master French coachbuilder Jean Daninos. Owners: F. Hobbs, O. T. Watton.***

FACEL VEGA

The cars called Facel Vegas were the creation of French master coachbuilder, Jean Daninos, who had established a good business making special bodies for cars such as Fords, Panhards and Simcas. The only trouble was that these were rather unexciting makes and, after gauging the reaction to early experiments shown at the Paris Salon in 1950, Daninos decided to build his own grand touring car. The results were to prove so successful that they would be compared with the products of Rolls-Royce.

HK500

YEARS MADE: 1958–62.
ENGINE: V8, 6286 cc, pushrods, overhead valves, 300 bhp at 4400 rpm.
FRAME: Tubular.
PERFORMANCE: 145 mph (233 km/h).
CURRENT PRICE GUIDELINE: ** Relatively rare, set to appreciate.
PROBLEMS TO LOOK FOR: Corrosion, neglected bodywork, non-originality.
CLASSIC CAR STATUS: *** Likely to improve in value because of elegant coachwork.

In the early 1950s Daninos did not have enough capital to make his own mechanical parts, so – his earlier Ford experiences in mind – he harnessed the lazy power and torque of an American engine to his own distinctively European chassis.

With a marvellous display of Gallic confidence, he shrugged off the facts that the French government frowned on anything imported and would require him to export five out of every eight of his foreign-engined cars, and that French taxation laws had all but killed off the indigenous luxury car industry. Daninos was sure that he could succeed where others had failed.

He chose the biggest and best contemporary American power unit: Chrysler's famous 180 bhp hemi-headed V8, along with its excellent Torqueflite automatic gearbox. The American practice of using only three gears in manual adaptations was unthinkable for a European machine, so a proprietary four-speed box made in France was offered, with a British Salisbury rear axle of unimpeachable quality. The same French gearwheel works at Pontà-Mousson, only a few miles from the old Bugatti factory, supplied a cam and roller steering box with equally traditional finned light-alloy drum brakes. It was unfortunate that these brakes did little to halt the progress of the first Facels, although this was not unusual in 1954, as fashionable new lines in all-enveloping coachwork shrouded the drums from a cooling airflow. The steering was heavy and vague; but so was that used on most contemporary cars. The tubular steel chassis was far from sophisticated, but by comparison with others it was quite good.

None of that mattered when the Facel Vega's bodywork was revealed. It was of pillarless construction and extraordinarily attractive. Despite its size – 15 ft (4572 mm) long and nearly 6 ft (1820 mm) wide – Chrysler's mighty Hemi could propel this very luxurious car to 120 mph (193 km/h) and provide startling acceleration. All France was gripped in a fever of excitement at the sight of such a magnificent car, when it emerged in 1955 from Daninos' factory in Colombes. But few Frenchmen were able to buy one: of necessity, most Facel Vegas were for export, and, with hindsight, it can be seen that the lack of a strong home market became a terrible handicap.

Like all great coachbuilt cars, each one was individually made, which meant that it was easy to change the specification. Thus, as more powerful Chrysler engines appeared, they soon found their way into Facel Vegas. The first cars became known, retrospectively, as FV1s, the FV2 having a 250 bhp V8 with a higher compression ratio as better petrol became generally available, on the way to the FV2B with a 5.4-litre engine giving 285 bhp. It was at this point that the overstressed brakes were fitted with a servo to help slow the very heavy cars.

It was natural that these beautifully built and very expensive cars would soon be compared with the products of Rolls-Royce. Daninos loved such adulation and showed his appreciation by stretching his chassis by 2 ft (600 mm) to make a four-door Facel Vega called the Excellence that was at least as big as a Rolls-Royce and a lot more elegant. When fitted with an even larger 6.4-litre V8 in 1956, it proved capable of 125 mph (201 km/h), making it one of the fastest big saloons in the world, although it remained one of the slowest to stop. The Excellence chassis was then shortened in 1958 to nearer the size of a FV2B to produce a new two-door pillarless coupé, the HK500. A dramatic reduction in weight eased the braking problem, provided that drivers did not attempt to use the top speed of 145 mph (233 km/h). This continuing problem was alleviated when disc brakes were fitted in 1960. The result was a boom year in which more than 200 Facel Vegas were built, still mostly for export.

Daninos then decided to attempt to crack the French market and this move led to his downfall. Building his own engine meant that development costs had to be spread over a far larger production run – and attempting to build a mass-production Facel Vega had bankrupted the firm by 1964.

SUMMARY

Plus points

Styling—more elegant than a Rolls-Royce, more attractive than contemporary American limousines, more practical than Italian luxury cars.

Dependable mechanical parts providing they are in good condition.

Minus points

High cost of restoring the body.

Difficulty in obtaining some mechanical parts.

Poor brakes on early models.

Snobbish rejection of Franco-American hybrid ancestry.

FERRARI

Ferrari has been the most consistent producer of classic cars since the firm was founded in 1946 by former Alfa Romeo exponent and owner of the Scuderia Ferrari racing team, Enzo Ferrari. Although the first Ferraris were pure racing cars, they are entitled to be called classics of the road because most of the competition which took place immediately after the Second World War did actually take place on the road: thus Ferrari sports racing cars made wonderful road machines!

212

YEARS MADE: 1951–2.
ENGINE: V12, 2562 cc, single overhead camshaft, 150 bhp at 6500 rpm.
FRAME: Tubular.
PERFORMANCE: Typically 123 mph (198 km/h), according to specification.
CURRENT PRICE GUIDELINE: *** Rare model set to appreciate due to historic styling and performance.
PROBLEMS TO LOOK FOR: Non-originality, mechanical deterioration.
CLASSIC CAR STATUS: **** Likely to improve in value as cars become older.

The earliest Ferraris had cycle-type wings which could be quickly detached to turn them into open-wheeled racers, but they soon began to acquire better streamlined full-width bodies for sports car events, with the single-seaters following a separate line of development. The basic running gear was very similar, however, with the first models taking the designation 166 after the cubic-centimetre capacity of one of their 12 cylinders. Interestingly, this first 2-litre engine, designed by Gioacchino Colombo, is still the basis for Ferrari's 12-cylinder cars of the 1980s, while the 166 was to develop into the classic 212 barchetta.

Colombo left Ferrari for rivals Maserati in 1950, however, and was replaced by his former assistant, Aurelio Lampredi, who designed a much larger V12 engine for the contemporary 4.5-litre Formula 1 racing car. Many of Ferrari's subsequent 'big block' V12-engined cars used this Lampredi engine.

Ferrari's chassis also remained little changed right into the 1980s, being of typical racing car-style tubular construction. But the bodies have varied widely as Ferrari has never had the facilities to build his own coachwork, preferring to leave that to outside specialists. The favourites in the early years were Pinin Farina, Vignale and Carrozzeria Touring, who produced the classic barchetta (or 'little boat') style in open and fixed-head forms when the 166 was enlarged through the 195 of 1950 to the 212 in 1951. Although these were essentially sports racing cars, it was evident that good profits could be made from touring cars, so the line was developed into the 3-litre 250 series in 1953 using the Lampredi engine and the more powerful, 4.1-litre 342.

▽ Touring's barchetta on Ferrari's 212 chassis – pure inspiration. Owners: J. Briggs, D. Nelson.

SUMMARY

Plus points
Superb styling.

Ideal entry in newly-popular historic rallies such as the Mille Miglia.

Sensuous performance.

Ready availability of parts—at a price.

Minus points
Very expensive to restore.

Needs racing-car-style sympathetic handling and maintenance.

410 SUPERFAST

YEARS MADE: 1956–60.
ENGINE: V12, 4963 cc, single overhead camshaft, 340 bhp at 6500 rpm.
FRAME: Tubular.
PERFORMANCE: Typically 168 mph (270 km/h), according to specification.
CURRENT PRICE GUIDELINE: * * * * Individual appeal of each car being different in an already exclusive series leads to high prices, although machinery is impractical to use.
PROBLEMS TO LOOK FOR: Non-originality, mechanical breakages, especially in transmission.
CLASSIC CAR STATUS: * * * * Value inflated by scarcity.

Although Ferrari sports racers were usually given more comfortable Pinin Farina touring bodywork, they still had a sufficiently strong performance to acquit themselves well in competition. But there were products from rival

△ ***Massive, but impractical, the Ferrari Superfast was really a styling exercise.***

manufacturers which could outrun the Type 342, so Ferrari responded with the far more powerful 375 in 1954, and this was subsequently developed into the Type 410 with a 4.9-litre version of Lampredi's original Grand prix engine. Before long, this very expensive model had acquired the name Superamerica because of its popularity with wealthy American customers.

These cars were produced only in very small quantities, but there were still fabulously wealthy people who wanted something even more exclusive, so Pinin Farina (which became Pininfarina in 1957) produced the occasional one-off model with the massive Lampredi engine installed in the smaller and lighter 250 chassis, and clothed in the most exotic coachwork. This was the Superfast, so called because it was capable, in theory, of 168 mph (270 km/h), although, in reality, the lightweight chassis and transmission were incapable of handling the full power. But who cared? The Superfast was purely for show.

SUMMARY

Plus points

Exclusivity.

Superb Grand Prix engine.

Minus points

Totally impractical to use because it was a styling exercise—thus dramatic lack of headroom and a transmission that wilts under power and torque of vast engine, plus handling that cannot cope with weight, aerodynamics and potential speed.

Potentially vast restoration costs.

250GT

YEARS MADE: 1959–61.
ENGINE: V12, 2953 cc, single overhead camshaft, 250 bhp at 7400 rpm.
FRAME: Tubular.
PERFORMANCE: Typically 157 mph (253 km/h), according to specification.
CURRENT PRICE GUIDELINE: **** Rare, appreciating fast in wake of 250GTO.
PROBLEMS TO LOOK FOR: Non-originality, mistreatment.
CLASSIC CAR STATUS: **** Will always be held back by 250GTO, a pity as in some ways it is a more usable car.

The original 250 series, based on the 166, 195 and 212, was a far more serious venture than the Superfast and led to the establishment of the 250GT line using a 3-litre version of the lighter and more compact Colombo engine, and a more advanced coil-spring front suspension than the old-fashioned leaf spring used before. These were, in effect, the first Ferraris to be produced in series to a standard specification. In most cases, they were fitted with bodywork by Pininfarina or Scaglietti, open versions being called Californias after their chief market.

It was during this time that races for production cars, rather than sports cars built purely for racing, became very popular, with the Ferrari 250GT as the most competitive car available. Other marques were catching up fast, and by 1959 Ferrari needed a lighter car, so an alternative short-wheelbase version was produced, clad in flimsy ultra-lightweight bodywork by Scaglietti. These models typically adopted the name berlinetta (berlina being a saloon, and berlinetta a small saloon). Meanwhile, Ferrari continued to dominate sports car racing with a series of models called Testa Rossa

('red head') after the colour of the cylinder heads in their Colombo engines. (Red traditionally denotes the most powerful product of an engine maker.) At this point, the 250GTs were fitted with slightly detuned versions of these all-out racing engines.

At the same time, the 250GT range continued to be expanded by the introduction of a new touring version with two small rear seats in the by then popular 'two-plus-two' configuration.

This car was further developed in 1964 with a 4-litre version of the Lampredi engine as the 330GT, while the two-seater cars were given a 3.3-litre Colombo engine for the 275 series.

These new two-seaters then became purely touring cars, marketed either as the 275GTB (for grand touring berlinetta) or as an open 275GTS (for grand touring spyder.)

These 275-series cars represented a considerable advance on the earlier

▽ ***The Ferrari 250GT short-wheelbase berlinetta lightweight, one of the most handsome GT cars.***

250s because they had a more sophisticated transmission in which the gearbox moved back to share the rear axle casing, allied with independent rear suspension. This helped the car's weight distribution so much that its handling and traction were greatly improved. Two years later, in 1966, the 275 series became even more formidable with a more powerful twin-overhead-camshaft version called the 275GTB/4 because it had, of course, a total of four camshafts. Further development of the 330 series in the same year yielded the 330GTC (for grand touring coupé) and 330GTS (grand touring spyder) using the 275's more advanced chassis with the larger Lampredi engine.

By 1969, the rival Italian marques, Maserati and Lamborghini, were hitting Ferrari road car sales, so the 330's capacity was increased still further to 4.4 litres for the 365GTC and GTS, and power was increased by a large margin in a 365GTC/4 variant from 1970.

SUMMARY

Plus points

Superb road-race performance, many events in which it can compete.

Beautiful styling.

Durable mechanics if treated well.

Good handling.

Tremendous driving satisfaction.

Ready availability of spares—at a price.

Minus points

Will always live in the shadow of the 250GTO, although it is more practical.

High costs of restoration.

Needs racing-car-style sympathetic handling and maintenance.

250GTO

YEARS MADE: 1962–4.
ENGINE: V12, 2953 cc, single overhead camshaft, 290 bhp at 7400 rpm.
FRAME: Tubular.
PERFORMANCE: Typically 165 mph (266 km/h), according to specification.
CURRENT PRICE GUIDELINE: *****+ Amazing rate of appreciation in past three years has taken this car into top bracket, where it will remain.
PROBLEMS TO LOOK FOR: Non-originality. Very high prices have encouraged manufacture of bogus cars.
CLASSIC CAR STATUS: ***** Has reached the top as the ultimate front-engined Ferrari competition grand touring car.

During the heyday of the 250GT, sports-car racing was becoming so fast that the machinery could frequently outperform Grand Prix cars. Consequently, the regulations were revised from 1962

▽ ***Ferrari's 250GTO won three world racing championships and has become one of the world's most valuable cars.***

△ ***Ferrari named his V6-engined cars after his son Dino, the 246GT becoming a favourite. Owner: John Swift.***

to make the world championship a competition for grand touring cars rather than pure racers like the Testa Rossa.

Ferrari rose to the challenge by developing the short-wheelbase light-weight berlinetta form of the 250GT into the ultimate variant of this line, the 250GTO (the 'O' standing for *omologato*, Italian for homologated, which meant that the car qualified for the new races). This machine, the fastest and most spectacular 250GT of all, was fitted with an uprated mechanical specification which included a Testa Rossa engine, and it also had a more aerodynamic nose. This revised front-end profile put performance before practicality and increased its top speed at the expense of cooling when it was standing still . . . but then nobody was seriously expected to go touring in the high-performance 250GTO, despite the rather misleading 'GT' part of its designation!

The less competition-minded customers were offered a more civilized version with an untemperamental engine and strong steel bodywork – rather than paper-thin aluminium – called the Lusso (for 'luxury'). It featured a shorter wheelbase together with a much more comfortable and – of course – luxurious interior with an exclusive price-tag to match.

The 250GTO remained as the last of a line when motor sport's governing body refused to homologate a mid-engined version called the 250LM (for Le Mans) in 1965, pointing out that it was too far removed from the basic production car to qualify.

SUMMARY

Plus points
- Exclusive appeal.
- Superb road-race performance, numerous events in which it can compete.
- Durable mechanics if treated well.
- Good handling.
- Tremendous driving satisfaction.
- Ready availability of spares—at a price.

Minus points
- Very high costs of restoration.
- Low-drag nose can make it temperamental at low speeds.
- Needs racing-car-style sympathetic handling and maintenance.

DINO 246GT

YEARS MADE: 1969–73.
ENGINE: V6, 2418 cc, twin overhead camshafts, 195 bhp at 7600 rpm.
FRAME: Tubular.
PERFORMANCE: 155 mph (249 km/h).
CURRENT PRICE GUIDELINE: * * * Set to appreciate as one of the prettiest modern Ferraris.
PROBLEMS TO LOOK FOR: Extensive corrosion, mechanical wear and mistreatment.
CLASSIC CAR STATUS: * * * * Delightful modern classic.

It was in 1967 that the first roadgoing Ferrari Dino appeared (earlier V6-cylinder racing cars having borne that name in memory of Enzo Ferrari's son Dino, who died in 1957 as the engine design was completed). The intention with this exceptionally pretty little mid-engined car was to compete in the lucrative 2-litre market established by Porsche. But it meant producing a car in much larger quantities than had previously been the case with Ferrari, so the aluminium body panels of the early Dinos – called 206GTs – were changed to the more economical steel in 1969, and the engine capacity was increased to 2.4 litres (as the 246GT) to cope with the resultant extra weight and to challenge the rapidly increasing performance of the Porsche 911S. A spyder version with a removable roof panel was then produced from 1971 onwards to compete with Porsche's 911 Targa.

SUMMARY

Plus points

Lovely styling.

Sweet handling.

Good performance for a relatively heavy touring car.

Ready availability of parts.

Minus points

Frequent body corrosion.

High costs of restoration.

Needs high standards of maintenance.

Brakes not up to modern standards.

365GTB/4

YEARS MADE: 1968–74.
ENGINE: V12, 4390 cc, twin overhead camshafts, 352 bhp at 7500 rpm.
FRAME: Tubular.
PERFORMANCE: 174 mph (280 km/h).
CURRENT PRICE GUIDELINE: * * *
Becoming rarer, and thus more valuable, as misguided buyers convert original fixed-head coupés into non-original spyder versions.
PROBLEMS TO LOOK FOR:
Mistreatment, mechanical and bodily deterioration.
CLASSIC CAR STATUS: * * * * Last and fastest of the great front-engined Ferrari GT cars.

At the other end of the range to the 246GT, great excitement was created by the introduction in 1969 of the 365GTB, a short-wheelbase 4.4-litre car with a top speed of 174 mph (280 km/h) that was to make it for years the fastest road car in the world. Everybody except Enzo Ferrari called it the Daytona after a great American race victory by his P4 prototypes; Ferrari proclaimed that it could not be called a Daytona because it did not fulfill all the criteria for a racing car.

Enzo Ferrari continued, however, to produce seemingly endless variations on his touring cars, including the 365GTC/4, which was made between 1971 and 1973, before being replaced by the 365GT/4 two-plus-two. This car then gave way in 1976 to the ultimate four-seater, the 400GT, with a 4.8-litre Lampredi-designed engine, and the option of the first automatic gearbox for a Ferrari.

▽ ***The fearsome 365GTB/4 Ferrari. Owner: The Patrick Collection.***

△ ***Italian bodymakers Pininfarina adapted the 246 Dino's lines to produce the Ferrari 308GTB. Owner: J. Marshall.***

SUMMARY

Plus points

- Tremendous engine.
- Immense performance.
- Reasonable handling.
- Brutal appeal of a bygone era.
- Ready availability of spares.

Minus points

- Handling and brakes more primitive than modern mid-engined Ferraris.
- High restoration costs.

308GTB

YEARS MADE: 1975–84.
ENGINE: V8, 2927 cc, twin overhead camshafts, 250 bhp at 7700 rpm.
FRAME: Tubular with central monocoque.
PERFORMANCE: 154 mph (246 km/h).
CURRENT PRICE GUIDELINE: * * *
Set to appreciate in value as cars become older.
PROBLEMS TO LOOK FOR: Mechanical wear, mistreatment, corrosion in later steel-bodied cars (early models had glass-fibre body panels).
CLASSIC CAR STATUS: * * Set to appreciate.

A 3-litre V8 version of the Daytona's engine was developed for a new Dino, the 308GT/4 two-plus-two, in 1973 – with somewhat unpopular styling by Bertone. Solving this styling problem, in 1975 Pininfarina designed what was to be a classic adaptation of the original 246GT shape – the 308GTB two-seater – which was followed by an open-topped GTS version in 1977.

These were all mid-engined cars, of course, as all Ferrari's racing cars had been since 1965, like Lamborghini's Miura and Countach production cars (see pages 97 and 100).

In 1984, true to its racing tradition, Ferrari built a limited-production competition version of the 308 called the GTO.

SUMMARY

Plus points

- Superb appearance.
- Excellent handling and performance.
- Easy to drive.
- Spares expensive but readily obtainable.

Minus points

- Cramped cabin.
- Noisy.
- Steel-bodied cars suffer from corrosion.

512BB

YEARS MADE: 1976–84.
ENGINE: Flat 12-cylinder, 4823 cc, single overhead camshafts, and 340 bhp at 6800 rpm.
FRAME: Tubular with central monocoque.
PERFORMANCE: 176 mph (283 km/h).
CURRENT PRICE GUIDELINE: * * *
Set to appreciate in value as cars become older.
PROBLEMS TO LOOK FOR: Mechanical wear, mistreatment, corrosion.
CLASSIC CAR STATUS: * * Set to appreciate.

The Daytona became the last of the great front-engined Ferrari GT cars when it was replaced in 1974 by the Berlinetta Boxer (so called because its 4.4-litre, Colombo-based engine had been flattened to save space, with the result that the pistons moved backwards and forwards towards each other like a pair of boxers' fists). Later, in 1976, the engine capacity was increased to 5 litres by lengthening the stroke and increasing the bore. This extra power made the 512BB, as the new car was called, smoother and easier to drive.

In recent years these engines have been uprated to fuel-injected *quattrovalvole* (four-valve-per-cylinder) form, with the two-plus-two bodystyle reappearing as the Mondial in 1981, and the 308 developing into the 328, while the 512BBi ('i' for injection) was replaced in 1984 by a restyled and more refined version, for which the name Testa Rossa was revived.

SUMMARY

Plus points
- Tremendous usable performance.
- Excellent handling.
- Attractive styling.
- Investment potential.
- Easier to drive than earlier and slightly faster 365GT4 BB.
- More attractive styling than later Ferrari Testa Rossa.

Minus points
- Expensive to run.
- Not so nimble or striking as some of its rivals.

FIAT

Few really small cars have ever achieved classic status, because the entire concept of the minimum car for a motorist of minimum means contradicts the thoroughbred status. In the case of the Austin Seven, its charm overcame many serious deficiencies, such as erratic handling and braking (see page 26). But in the case of another great small car, the Fiat 500 of 1936 – nicknamed 'Topolino', the Italian name for Walt Disney's irrepressible character Mickey Mouse – technical sophistication made it attractive to those who could afford cars of far greater size and cost, as well as to people who could manage only a minimal outlay.

△ *The Ferrari 512 Berlinetta, called the Boxer because its pistons work in opposition. Owner: L. Page.*

TOPOLINO

YEARS MADE: 1936–48.
ENGINE: In-line four-cylinder, 569 cc, side valves, 13 bhp at 3800 rpm.
FRAME: Twin girders, cross-braced.
PERFORMANCE: 55 mph (86 km/h).
CURRENT PRICE GUIDELINE: *
Fairly rare, due to age, excellent examples fetch more.
PROBLEMS TO LOOK FOR: Old age, corrosion and non-originality.
CLASSIC CAR STATUS: * * * Historically very advanced small car, set to appreciate soon.

The Topolino was the personal conception of Senator Giovanni Agnelli, who founded F.I.A.T. – the Fabbrica Italiana Automobili Torino – in 1899 and turned it into an industrial manufacturing colossus embracing not only cars but also commercial vehicles, ball bearings, shipbuilding, aero-engines, marine engines, aircraft, and railway rolling stock. Sidelines included financing Ferrari's Grand Prix racing 'for the good of Italy'.

With the success of the Austin Seven in mind, in 1933 Agnelli called together engineering staff from all his divisions and told them of his idea to produce a car costing only 5000 lire at a time when his cheapest product, the 508 Balilla, was a tremendous success at more than twice the price.

Soon a brilliant young designer from

the aviation section, Dante Giacosa, had produced clay models of what he thought the next Fiat small car should look like. Significantly, these models showed strong evidence of wind-tunnel work in that they were beautifully streamlined in an era when the average car displayed all the aerodynamic qualities of a house brick. Clever juggling of basic components, such as the engine which was set, aircraft-style, ahead of the radiator, enabled Giacosa to achieve the rounded nose vital for good air penetration without making the car unnecessarily long and heavy. He also, sensibly, resisted any temptation to offer more than two seats, so that the comfort of full-sized adults did not have to be compromised in what had to be a very small car.

Giacosa would have liked to have used front-wheel drive to give more space in the passenger compartment, but it proved impractical economically to manufacture the constant-velocity joints in the drive shafts that would be vital to such a system, so he had to settle for a conventional rear-wheel drive. However, thanks to the use of wishbone independent front suspension (which would have been needed with front-wheel drive) and the help of Fiat's chief car designer, Antonio Fessia, excellent handling was achieved. Other advanced features included hydraulic brakes and a 12-volt electrical system. The side-valve four-cylinder engine of only 569 cc and four-speed gearbox were quite conventional, but so economical and durable that the Topolino could be driven flat out at 55 mph (86 km/h) with the stunning fuel consumption of 55 mpg (5.14 litres/100 km) from its introduction in 1936. The credit for such speed and economy was largely due, of course, to the excellent streamlining but the greatest achievement in the design of the Topolino was in the supremely harmonious and original, yet logical, way in which the elements were combined.

The Topolino was successful beyond even Giacosa's dreams as it not only sold to the masses, but also competed successfully in competitions as diverse as the Monte Carlo Rally and the Mille Miglia, besides crossing deserts and climbing mountains . . . all the feats which would have seemed beyond a lesser character than Mickey Mouse!

Despite the intrusion of the Second World War, production had reached 122,000 by the time the Topolino was uprated with an overhead-valve engine as the 500B in 1948, followed by the 500C in 1949 which went on to the 376,000 mark taking the total number of Topolinos past half a million by 1955.

The modernized 600 which replaced the Topolino had a rear-mounted engine to liberate extra seating space, and was just as much of a sensation. By 1957 Fiat was able to use the 600 as a base for the Nuova 500, which became the direct replacement for the Topolino.

SUMMARY

Plus points

Brilliant design.

Excellent handling.

Cheeky performance and appearance.

Easy to drive.

Practical to run once in good condition.

Minus points

Relatively sedate performance by modern standards.

Cramped cabin.

▽ ***Fiat's dainty little Topolino, the creation of one of the greatest car designers, Dante Giacosa.***

FORD

Ford established an enviable reputation for producing the cheapest and most reliable cars during the early days of motoring – and then found that a far more glamorous image was needed in the early 1960s to attract youngsters who had a lot more money to spend. The giant car manufacturer realized that if it was to survive as one of the greatest mass-producers, it would have to adopt a vibrant new image. So it tried first to take over the greatest high-performance name of all, Ferrari, but – like many others before – found it impossible to tie down the autocratic Enzo Ferrari, who chose instead to accept finance from Fiat, because his firm 'was a national treasure which should not be allowed to leave Italy'! Ford then attacked the competition world from every angle, eventually settling for a heavy investment in Colin Chapman, of Lotus, the only man that Ferrari recognized as his equal. Finally, Ford stimulated competition among its own management to promote the new sporting image.

MUSTANG

YEARS MADE: 1964–73.
ENGINE: Typically V8, 4736 cc, overhead valves, pushrods, 240 bhp at 5750 rpm.
FRAME: Integral.
PERFORMANCE: 127 mph (204 km/h).
CURRENT PRICE GUIDELINE: *
Getting rarer, will appreciate as many cars depreciate.
PROBLEMS TO LOOK FOR: Corrosion, mistreatment, lost items of trim.
CLASSIC CAR STATUS: ** Trend-setting pony car, the first really popular Ford sports car.

By the early 1960s Ford was seeking a foothold in the young car-buyer's market, and sales director Lee Iacocca came up trumps with a car which looked sporty, carried four people, yet cost less than $2500 because its mechanical components were shared with saloon cars produced in huge quantities. Much of his inspiration came from the first Ford Thunderbird, a relatively small car which had been built to compete with General Motors' Chevrolet Corvette in the opulent 1950s (see page 44). Like the early Thunderbird, the new model was called a 'personal car' because it was available with so many options that it could be made to look different from the one next door. This meant that owners of the bottom-of-the-range 170 cu in (2.8-litre) six-cylinder cars could boast of economy, while top-liners with a 289 cu in (4.7-litre) model could demonstrate a searing performance, with many variants of trim and mechanical specification.

Paradoxically, the new generation was also locked into nostalgia for the freedom of the Wild West that it had seen only in Hollywood movies, so it was hardly surprising that Ford's new car was named after the cowboys' horse: the Mustang.

Never has a car satisfied its target market more. The Mustang sold 22,000 on its first day in 1964, and more than 400,000 in its first year of production. The opposition had no choice but to follow suit – General Motors with the Camaro and Firebird (named after the Indian side in cowboy games), leaving American Motors to opt for 'high-school' action with the Javelin, while Chrysler cast around for what was left in leisure activities with the Barracuda. But the young at heart had the best name of all for such models: pony cars.

They had all the brawn of muscle

△ ***American dream: the Ford Mustang sold 22,000 on its first day. Owner: Bob Hodges, 'Yesterday's Wheels', Hartford City Ind.***

cars like the massive Pontiac GTO and rumbling Dodge Charger (see page 59), with a hint of the athletic sophistication of smaller machines. Such basic desires were fuelled by a competition programme organized by Texan racing driver Carroll Shelby, who not only dressed like a cowboy, but fought like one too, in Ford's opinion. His AC Cobra (see page 9), with the same V8 engine as the top-line Mustang, toppled Ferrari's 250GTO (see page 68).

SUMMARY

Plus points
Excellent usable performance.

Very durable and reliable.

Excellent availability of spares.

Nostalgic 1960s styling.

Minus points
Extensive corrosion in many examples.

Outdated plastic interior not yet back in fashion.

GT40

YEARS MADE: 1964–6.
ENGINE: V8, overhead valves, pushrods, typically 4736 cc, 325 bhp at 5750 rpm.
FRAME: Steel monocoque.
PERFORMANCE: Typically 184 mph (296 km/h).
CURRENT PRICE GUIDELINE: **** Rapid appreciation as the last of the great roadgoing sports racing cars.
PROBLEMS TO LOOK FOR: Non-originality, extensive corrosion in non-rustproofed, lightweight steel monocoque.
CLASSIC CAR STATUS: ***** Recent rise in values has taken it to the top.

At the same time as Shelby was performing his miracles with the Cobra and Mustang, Ford was also using Colin Chapman to do a similar job with its components in international single-seater and saloon car events. Unfortunately Ford was to discover that – because his Lotus cars were already so well known – Chapman's firm took much of the glory. So they turned to one of his rivals in sports-car racing, Eric Broadley of Lola, to help launch an advanced new mid-engined Ford that could beat Ferrari on its own ground in the world's most famous endurance race at Le Mans.

The resulting GT40 – so named because it was a prototype GT car only 40 in (1016 mm) high – became much more than a classic Le Mans winner. It was the last great sports racer that could also be used on the road. Everything that followed it was impractical for such purposes, fitted with ever-widening tyres which had to be changed in rain or would not last in the dry.

By 1966, enough GT40s (powered by what were essentially the same engines as those used in the Mustang and AC Cobra) had been produced for the model to be homologated for GT racing. Ferrari fought back to humble Ford by winning 1-2-3 on the American company's home ground at Daytona early in 1967, so Ford responded by uprating the GT40 so that it became the first car to exceed 200 mph (322 km/h) on the Mulsanne straight in one of four historic Le Mans wins which helped make the model an all-time classic.

▽ ***The GT40 was the Ford which beat Ferrari: with its derivatives, it won at Le Mans a historic four times.***

SUMMARY

Plus points
Stunning performance.

Relative durability except for corrosion.

Ready availability of spares.

Minus points
Cramped, spartan cabin typical of a sports racer.

Rugged ride.

Bad ventilation.

Monocoque prone to corrosion.

Performance can be beyond the capabilities of average driver.

FRAZER NASH

With the exception of the occasional saloon car, or Formula 2 racing machine, the British marque Frazer Nash concentrated exclusively on sports cars. Initially, these were the creation of racing driver Captain Archie Frazer-Nash, who used a hyphen in his name, although the cars did not.

LE MANS REPLICA

YEARS MADE: 1950–3.
ENGINE: In-line six-cylinder, 1971 cc, overhead valves, pushrods, typically 120 bph at 5500 rpm.
FRAME: Tubular.
PERFORMANCE: 110 mph (177 km/h).
CURRENT PRICE GUIDELINE: * * *
Relatively rare, but a very high percentage of survivors are in excellent condition.
PROBLEMS TO LOOK FOR: Possible non-originality.
CLASSIC CAR STATUS: * * Ought to be far higher, marque under-publicized.

The first Frazer Nashes were developed in 1924 from the G.N. cycle cars (so called because they were very rapid motor cycle-engined machines) which Archie had built in cooperation with another engineer, H. R. Godfrey – hence the initials, G.N. The early cars consequently featured a similar, motor cycle-derived chain-drive transmission to that used in the G.N.s. Because ratios could be changed very quickly in this rather noisy, but very efficient, 'dog clutch' transmission, and since the spartan Frazer Nashes were very light, they were soon renowned as some of the fastest vintage sports cars.

Typical features of early examples were a proprietary side-valve Anzani engine and polished aluminium bodywork, and there was an ultra-lightweight Boulogne model for the really serious racing man.

The Anzani-powered cars were followed by numerous Meadows-powered models, of which the three-seater Tourist Trophy Replica was the most popular. As soon as the performance of other cars looked like catching up with that of a 'Nash – as the Frazer Nash became known – a new model was introduced that went even faster, culminating in the two-stage supercharged Shelsley model and the single-overhead-camshaft Gough-engined cars. Later machines were fitted with the smoother and more refined twin-overhead-camshaft six-cylinder Blackburne engine.

Late in 1928, Frazer-Nash, who had always been the driving force behind the marque, became ill and the firm had to be saved by car dealer H. J. Aldington's company, AFN. Aldington proved to be a far better businessman than the founder.

It was clear that the primitive chain-drive transmission could not survive long, no matter how efficient it might be, so by 1934 Aldington was importing the German BMW cars that had been proving the Frazer Nash's equal in European rallies (see page 34). The chain-driven cars continued to be made in small quantities, however, until the Second World War, sometimes with a BMW 319 engine.

Meanwhile, Aldington had linked up with the Bristol aircraft company (see page 38) and this resulted in the later, and far more powerful, BMW 328 engine going into production as a Bristol car unit. This was to form the basis of a completely new Frazer Nash sports car with a tubular chassis and independent front suspension. Like the 'Nashes of old, it was exceptionally light with excellent roadholding.

The new car was introduced in 1948 as the High-Speed Model with the

▽ The Frazer Nash Le Mans Replica, one of the most successful postwar British competition cars.

minimum of bodywork and with cycle-type wings, as on many contemporary dual-purpose road-and-racing cars.

An early success in the Le Mans 24-hour race the following year resulted in the High-Speed Model being renamed the Le Mans Replica, and the car was later fitted with de Dion rear suspension as the Competition Model, resulting in even better handling.

In 1952 a Frazer Nash won the American Sebring 12-hour race, leading to the introduction of a new model called the Sebring. This had strikingly pretty, all-enveloping bodywork, since the old cycle-type wings had been outlawed for international competition. There was also a Targa Florio 'fast touring model'.

By 1955, however, the Frazer Nash had reached its zenith with a 150 bhp model for the Tourist Trophy race. No more power could be extracted from what was already a very expensive engine, and further prototypes were fitted with an Austin-Healey 100 or an Armstrong Siddeley Sapphire unit in an attempt to reduce costs.

When BMW began to recover from the war and made a 2.6-litre V8 engine, this was also tried in Frazer Nash's best chassis, that of the de Dion-axled Competition Model. This two-seater coupé, the Continental, was an attempt to enter the market for very exclusive cars, dominated by Ferrari.

However, the Continental cost four times as much as a Jaguar XK150, which offered a similar performance, and – like the Allards produced nearby in London (see page 16) – it priced itself out of the market: production ceased in 1957. Despite this, Aldington's firm, AFN, went from strength to strength, as he had taken on the British concession for the increasingly successful German marque – Porsche.

SUMMARY

Plus points

Excellent usable performance.

Superb handling.

Durability.

Tremendous driving appeal.

Minus points

Expensive to restore.

△ ***Hispano's Alfonso, called after the Spanish king, became the world's first sports car. Owner: S. F. Mitchell.***

HISPANO-SUIZA

Some of the most extraordinary and glamorous cars ever made bore the name of Hispano-Suiza – denoting the combination of the Swiss brains of engineer Marc Birkigt and Spanish finance at their headquarters in Barcelona. It is odd to recall, therefore, that Hispano-Suiza is best remembered as a French marque!

The Hispano-Suiza story began when Birkigt joined a Spanish electrical company in 1900 and subsequently designed a car which was first called La Cuadra and then the Castro. This model became the first Hispano-Suiza when the firm was refinanced in 1904.

ALFONSO

YEARS MADE: 1912–14.
ENGINE: In-line four-cylinder, 3619 cc, side valve, 63 bhp at 2300 rpm.
FRAME: Twin girder, cross-braced.
PERFORMANCE: 80 mph (129 km/h).
CURRENT PRICE GUIDELINE: *** Undervalued on a historical basis.
PROBLEMS TO LOOK FOR: Non-originality, simple old age.
CLASSIC CAR STATUS: *** Ought to be higher as it was effectively the world's first sports car.

Happily for Hispano-Suiza, Spain's King Alfonso XIII was a motoring enthusiast and he soon became a regular customer. His English wife, Ena, bought him a two-seater Hispano-Suiza for his 23rd birthday in 1909 and in the following year Hispano proudly called its latest competition model, the 15T, the Alfonso. This car, which was far lighter than many contemporary racers, was then adapted for road use, going into production in 1912.

In reality, the Alfonso was the world's first sports car because it was an absolute joy to drive compared to its competition, which invariably weighed around twice as much. Because it was so light, Birkigt was able to use a relatively rigid chassis with soft springs which gave the Alfonso a ride more like that of a modern motor car and agility that was quite extraordinary for its time. The engine, which had a great deal of pulling power because of its long stroke, made the car even easier to drive, with gears which were so well machined that they were exceptionally smooth to engage . . . at a time when crunching changes were the norm.

SUMMARY

Plus points

Superb performance for its year.

Excellent nimble handling.

Beautifully made.

Durability.

Minus points

None other than scarcity!

H6C BOULOGNE

YEARS MADE: 1924–9.
ENGINE: In-line six-cylinder, 8021 cc, single overhead camshaft, 160 bhp at 3000 rpm.
FRAME: Twin girders, cross-braced.
PERFORMANCE: 110 mph (177 km/h).
CURRENT PRICE GUIDELINE: * * * * Far better value than many contemporaries.
PROBLEMS TO LOOK FOR: Non-originality, old age.
CLASSIC CAR STATUS: * * * * Would be even higher if it did not live in the shadow of the Hispano V12.

The only problem facing Hispano-Suiza in the early days was that not enough Spaniards could afford its products, so a satellite factory was established in Paris when the Alfonso was launched. From 1914, this works produced a range of V8 aluminium aero-engines designed by Birkigt which did much to win the First World War.

When the war was over, one bank of a prototype V12 aero-engine of advanced single-overhead-camshaft design was used for the power unit in a new 6.6-litre Hispano-Suiza luxury car, the H6B. In the best traditions of aero-engines, this unit was not only very light for its size, but also very reliable. Like the Alfonso, the H6B had a light, yet rigid, chassis. With so much power and torque on tap, you could drive an H6B from 8 to 86 mph (13 to 138 km/h) in top gear, a performance which fitted the model ideally for the luxury car trade. But, unlike the Alfonso, you could stop this powerhouse with ease, such was the efficiency of the servo-assisted brakes on all four wheels. It needed only the introduction in 1924 of a 110 mph (177 km/h) 8-litre short-chassis sports model, the H6C Boulogne, to bolster the marque's reputation even further.

SUMMARY

Plus points
Superb build quality.
Excellent performance.
Durability.
Easy to drive.

Minus points
Bulky.
Expensive to restore.

▽ ***Hispano's top sports model of 1924 was named the Boulogne. Owner: Lt.-Col. Kerslake.***

TYPE 68bis

YEARS MADE: 1934–8.
ENGINE: V12, 11,310 cc, single central camshaft, pushrods, 250 bhp at 3000 rpm.
FRAME: Twin girders, cross-braced.
PERFORMANCE: 108 mph (174 km/h).
CURRENT PRICE GUIDELINE: ***** Undervalued compared to a Bugatti Royale.
PROBLEMS TO LOOK FOR: Non-originality, particularly of coachwork.
CLASSIC CAR STATUS: ***** One of the ultimate acquisitions.

While the French branch of Hispano-Suiza based its production on luxury cars, the Spanish factory concentrated on the more mundane machines and commercial vehicles better suited to its home market. When the Wall Street crash of 1929 threatened Hispano-Suiza's luxury car sales, the company took over the French firm Ballot to produce a smaller, 4.5-litre vehicle for a while. Then, with magnificent contempt for the Depression, Hispano-Suiza introduced its greatest car – the Type 68. This model was fitted with a giant V12 engine of 9.5 litres to give it extra pulling power, as it was bound to be endowed with some magnificent, but weighty, bodywork. The engine's overhead camshaft configuration – although far from noisy – was abandoned, however, as Hispano-Suiza wanted to emulate the extremely quiet running of the cars from its British rivals, Rolls-Royce (see pages 148-53).

But despite the weight of extravagant bodywork, the engine was so flexible and powerful that the Type 68 had a better performance than many of the best sports cars of the day. As if that was not enough, the capacity was then increased to 11,310 cc for even more pulling power in the Type 68bis. Comparison with Bugatti's Royale (see page 40) was inevitable. The big difference was that Hispano-Suiza managed to sell no less than 110 of these very expensive machines against Bugatti's three. However, as with the Royale's engine, the Type 68's unit also found its way into French railcars.

Production of the Type 68bis eventually came to an end when Hispano-Suiza became increasingly involved in armaments manufacture. Then the factory in Barcelona was taken over by the Pegaso truck firm which was to produce – as well as buses and lorries – one of the most exotic post-war grand touring cars (see page 140). In France, the Parisian arm of Hispano-Suiza merged with Bugatti to become an important part of the French aerospace industry.

△ ***Hispano's greatest luxury car, the Type 68bis. Owner: C. W. Hampton Sussex.***

SUMMARY

Plus points
- Magnificent design.
- Superb build quality.
- Excellent performance.
- Durability.
- Even easier to drive than an H6!

Minus points
- Bulky.
- Expensive to restore.

H.R.G.

Following the demise of the G.N. car firm run by H. R. Godfrey and Captain Archie Frazer-Nash, Godfrey started a business repairing G.N. cars while Frazer-Nash evolved his own sports cars (see page 76). Then, as AFN concentrated on its new BMW imports, Godfrey saw a continuing market for a traditional sports car like the Frazer Nash. So he went into business in 1935 with two other engineers, called Halford and Robins, to build a similar car – the H.R.G.

H.R.G. 1500

YEARS MADE: 1936–56.
ENGINE: In-line four-cylinder, typically 1496 cc, single overhead camshaft, 61 bhp at 4800 rpm.
FRAME: Twin girders, cross-braced.
PERFORMANCE: Typically 83 mph (134 km/h).
CURRENT PRICE GUIDELINE: * * Relatively rare, but many good examples survive.
PROBLEMS TO LOOK FOR: Deterioration through age.
CLASSIC CAR STATUS: * * Undervalued at present.

The H.R.G. shared the same type of very stiff suspension as Frazer-Nash's creations: quarter-elliptic front springs and a tubular axle located by the shock absorber arms. However, even Godfrey realized that the chain-drive transmission had had its day, and he substituted a conventional Moss four-speed gearbox and live rear axle suspended by half-elliptic springs. The simple two-seater bodywork followed traditional lines: it was narrow, draughty, and equipped with an exposed slab tank at the rear, and it came without such luxuries as a heater or glass side-screens – but it was very light. To complete the Frazer Nash-like illusion, it had a twin-carburettor Meadows four-cylinder 1.5-litre engine. From 1938, however, when supplies of the Meadows engine dried up, a similar-sized unit from the quantity-production Singer car was used.

There were enough drivers able to afford what was essentially a cheaper car than a Frazer Nash to keep H.R.G. in business until after the Second World War, especially since the company also undertook general engineering work, including a speciality of cylinder head development. It was then quickly back into automotive action in 1946 with a slabsided 'aerodynamic' model using the same chassis and running gear. This new car was by no means a success since, as well as being rather ugly, the wide and rather flimsy lightweight body seemed all too likely to part company with the still very stiffly sprung chassis.

Consequently, H.R.G. concentrated on its basic pre-war model, now fitted with single-overhead-camshaft Singer engines of 1100 or 1500 cc capacity to qualify it for the popular classes in the international sports car races and rallies where the 'chain gang' Frazer Nashes had starred. These engines were stripped down in the H.R.G. works after delivery from Singer and modified to suit the customers' – and Godfrey's – tastes, with a higher compression ratio, revised manifolding, and a shortened stroke to bring them within their class capacity limit. This work raised the power output considerably, to 61 bhp for the 1500 engine and 44 for the smaller unit. Hardly anything else altered, though, including the old-fashioned cable-operated brakes.

Because of its lightness, the H.R.G. remained a highly competitive sports car, with annual sales peaking at 70 in 1946 and 1947 in a home market starved of new cars. By 1949, however, when competitors were starting to come out with more modern designs that were far less tiring to drive, annual production had fallen to 14, and subse-

quent H.R.G.s were made only to special order. The 1100 cc car was dropped in 1950, but in 1955 Singer's new SM1500 engine was adopted with similar modifications that brought its capacity down to just under 1500 cc. In this form it gave 66 bhp and a slightly better performance.

H.R.G.'s engineering work included developing an advanced new twin-overhead-camshaft cylinder head for the SM engine, which was subsequently used in cheaper cast-iron form for Singer's Hunter 75 saloon.

At the same time, the company was developing an extremely advanced new sports car with a very stiff tubular frame – quite unlike the previous flexible affair – and independent wishbone suspension front and rear by the popular transverse leaf springs. It also had magnesium alloy wheels and disc brakes all round, at a time when these had been used on only the most exotic sports racing cars, such as the D-type Jaguar (see page 87). Naturally this car, equipped with modern alloy bodywork which bore some resemblance to John Tojeiro's AC Ace (see page 9), and powered by a tuned version of H.R.G.'s twin-cam Singer-based engine, was seen as a formidable competition machine.

However, only two had been made before Singer was taken over, in 1955, by the Rootes Group, who decided against putting the model into production and even discarded the saloon car's twin-cam cylinder head. H.R.G. could not afford to put what would have been a fairly expensive model into production, and assembled the last of its old-style 1500 cc cars in 1956 before continuing with general engineering work. Nevertheless, it could be said that H.R.G. had effectively extended the life of the pre-war Frazer Nash by some 20 years!

SUMMARY

Plus points

- Charming pre-war nostalgia.
- Superb appearance.
- Good handling.

Minus points

- Very uncomfortable ride – can shake itself to pieces.
- Scarcity of spares.

▽ ***The traditional lines of the H.R.G. – in 1100 cc or 1500 cc form – always stayed with the marque.***

ISO GRIFO

The Italian Giorgetto Giugiaro has become one of the most significant car designers of all time. A string of Giugiaro-styled small cars was introduced in the 1970s and 1980s – but he really made his name with extraordinary cars like the Iso Grifo. Even its name was fabulous, alluding as it did to the griffin – a creature with an eagle's head and wings and a lion's body. It will be remembered that the griffin was said to be the particular enemy of horses . . . and, of course, Iso's Italian rival, Ferrari, bore the badge of a rampant horse!

The Grifo, produced by the Iso refrigerator company run by Renzo Rivolta, was a true work of art. It had excellent handling thanks to Giotto Bizzarrini, previously famous for the legendary Ferrari 250GTO (see page 68), and an abundance of smooth and reliable power from a Chevrolet Corvette V8 which was unfettered by export emission controls (see page 46). In many ways the Iso Grifo was one step ahead of the 250GTO in that it was built on a steel platform chassis on to which numerous small steel body panels were grafted. Nuccio Bertone supplied Rivolta with the dies for these body panels and took the lion's share of the credit for the car's fabulous lines: few people outside Italy had heard of Giugiaro at that time.

ISO GRIFO

YEARS MADE: 1963–73.
ENGINE: V8, typically 5359 cc, overhead valves, pushrods, 385 bhp at 5000 rpm.
FRAME: Steel platform.
PERFORMANCE: 160 mph (257 km/h).
CURRENT PRICE GUIDELINE: * * * Undervalued by comparison with V12-engined rivals.
PROBLEMS TO LOOK FOR: Possible corrosion in chassis, general deterioration.
CLASSIC CAR STATUS: * * * Set to rise higher on superb styling and performance.

In 1959, the 20-year-old Giugiaro, an artist's son, was hired by Bertone to work in his Turin studio. The young man started by designing the Alfa Romeo 2000 Sprint and had completed the Giulia GT and Gordon Keeble in 1960, before he was even 22 years old. A singularly beautiful one-off body on a Ferrari 250GT followed (see page 66).

In 1962, a car called the Iso Rivolta went into production. It looked remarkably like the still-born Gordon Keeble, but this resemblance was justified because every great artist needs the satisfaction of seeing his best work reach the market for which it was intended. It was ironic, therefore, that the Gordon Keeble project was refinanced and production of the British GT car – still difficult to distinguish from an Iso Rivolta – started in 1964!

Meanwhile, Bertone had 'lifted' the other still-born Giugiaro design – that of the Ferrari 250GT project – for Iso's next car, the Grifo of 1963. To add to the confusion, the same basic body

design was used by Bizzarrini (a consultant on the Grifo project) for his own GT Strada 5300. Nevertheless, the Grifo survives as the classic version, since it was first into production.

Early examples had a 5.4-litre engine tuned to give 365 bhp with a top speed of 160 mph (257 km/h), which was more than a match for the 330GT competition from Ferrari (see page 68). The Grifo was not particularly fast off the line, however. Its 0 to 60 mph (97 km/h) time was only 7.4 seconds against, for instance, the Aston Martin DB4's figure of less than 7 seconds (see page 22). This was because the Grifo had an extra-high first gear ratio, the wisdom of which can be appreciated when you learn that a Grifo could exceed 60 mph (97 km/h) in first gear (ideal for getting clear of clogged Fiats), 10-80 mph (16-129 km/h) in second (for mountain roads, tight lanes and mobile chicanes on the *autostrade*), 20-100 mph (32-161 km/h) in third (for sweeping bends), and an optimum 100 mph (161 km/h) in top at less than 4000 rpm. At this speed you could still leave any rival supercar behind with no more than a nudge on the accelerator. The 7-litre Grifo introduced in 1968 was even faster, and its competition version hit 186 mph (299 km/h) at Le Mans.

The reason for paying three times the price of a Jaguar E-type, which could go nearly as fast as a normal Grifo (see page 89), was that you were buying a beautifully built grand touring car, and one that could hold its own with anything on the road, but in an infinitely more elegant manner. Giugiaro's lines remain the classic inspiration for any front-engined GT car.

In stark contrast to many supercars, the interior was almost as timeless as the rest of the body because Giugario was able to avoid gimmicks. Thus the seats were plain and simple, yet functional in that they gave better and less obtrusive support than many more elaborate designs. It was also hard to see how the comprehensive array of instruments could be better arranged than they were – grouped neatly, and not too close, in front of the driver.

◁ The superb lines of the Iso Grifo were an early example of stylist Giorgetto Giugiaro's work.

The 1973 oil crisis dealt Iso a mortal blow, but by then some 400 Grifos had been built, with some of the later ones being powered by Ford V8 engines. The majority of Grifos were fixed-head coupés, although there were a few open examples which were as graceful as Giugiaro's later Maserati Ghibli (see page 114). In fact, this styling was carried over to the smaller, and very pretty, Ferrari-engined Fiat Dino sports car which was to be Giugiaro's last great work for Bertone.

It just went to show what Ferrari had missed when it failed to put the grand tourer into full-scale production!

SUMMARY

Plus points

Marvellous appearance.

Excellent performance.

Easy to drive.

Very durable mechanics.

Well-made body.

Ready availability and low cost of mechanical spares.

Minus points

Potentially high cost of body restoration.

Possibility of corrosion in floorpan.

JAGUAR

Every Jaguar ever made, with the exception of the most recent XJ6 launched in 1986, was in essence the inspiration of one man, William Lyons, who in 1956 received a knighthood for his services to the British motor industry.

Even the current XJ6 saloon and its partner, the XJ-S, bear the stamp of Lyons, the master stylist, because they follow on so faithfully from the shapes he created.

Lyons was not only one of the most gifted stylists of the 20th century, he was also an extremely talented businessman who created one of Britain's most successful motor manufacturers from early beginnings tucked away in the back streets of the English seaside resort, Blackpool, in 1920 with a partner, William Walmsley.

S.S. 100

YEARS MADE: 1936–46.
ENGINE: In-line six-cylinder, 3485 cc, overhead valves, pushrods, 125 bhp at 4500 rpm.
FRAME: Twin girders, cross-braced.
PERFORMANCE: 101 mph (163 km/h).
CURRENT PRICE GUIDELINE: * * * Relatively rare, but many surviving examples now in good condition.
PROBLEMS TO LOOK FOR: Deterioration through age.
CLASSIC CAR STATUS: * * Underrated as one of the most beautiful pre-war sports cars.

Lyons started by building motor cycle sidecars which were called Swallows to reflect their graceful line. With the advent of the Austin Seven in 1922 (see page 26), the company began to concentrate more on motor cars, partly because sidecar work tended to be seasonal and partly because building special bodies for small cars like the Austin Seven was more profitable.

These bodies not only looked good but were also extraordinarily inexpensive due to the partners' amazing talent for keeping costs down. In fact, although a Swallow-bodied car cost more than the standard product, it was considerably cheaper than coachbuilt machines on other chassis. Swallow's products became popular during the economic depression which gripped Britain in the late 1920s and early 1930s, as people 'traded down' from more expensive cars, but balked at being seen driving a common Austin Seven!

It was during this era, in 1928, that the high overheads of transporting chassis from Midland car-making centres to Blackpool, in the North, and a shortage of skilled labour locally, forced Swallow to move to Coventry. Gradually the scope of its operation widened, to Standard chassis as well as the popular Austins, before Swallow became a car manufacturer in its own right in 1931, using mechanical parts made specially by Standard.

The first of these models was called the S.S.1 (it has not been revealed whether these initials stood for Standard-Swallow, Swallow-Standard, Standard Special or Swallow Special). This long, low, rakish six-cylinder coupé

was an outstanding bargain at £300 and scored a spectacular sales success. However, customers were already beginning to grumble that these cars looked far faster than they actually were, so the S.S.1 was soon being fitted with larger and more powerful engines, while a smaller S.S.II was introduced in 1934 to make sure that the original target market was catered for. The operation was becoming too big for Walmsley, however, so he resigned, leaving Lyons in sole charge.

One of Lyons' first moves was to create an engineering department to develop the company's own mechanical parts, at the same time hiring the brilliant freelance engineer, Harry Weslake, to extract a great deal more power from the basic 2.6-litre Standard engine. In 1935, the revised unit was installed in a special lightweight two-seater version of the S.S.1, called the S.S.90 because it could easily exceed 90 mph (145 km/h). This beautifully styled sports car was to be the forerunner of one of the most spectacular machines made before the Second World War. Fitted in 1936 with an overhead-valve 3.5-litre engine, with its power output boosted from 70 bhp to 105, the new sports car attained the magic 100 mph (161 km/h) and was proudly titled the S.S.100. It was to be the most glamorous model in a whole new range of sporting cars powered by overhead-valve engines. This new breed of cars would be called Jaguars, named evocatively after an animal which was renowned for its qualities of grace and speed.

SUMMARY

Plus points

Superb styling.

Durable mechanics.

Safe handling.

Good performance.

Minus points

Relatively expensive to restore.

XK120

YEARS MADE: 1948–54.
ENGINE: In-line six-cylinder, 3442 cc, twin overhead camshafts, typically 160 bhp at 5300 rpm.
FRAME: Twin girders, cross-braced.
PERFORMANCE: 125 mph (201 km/h).
CURRENT PRICE GUIDELINE: * * * Undervalued compared to contemporary opposition.
PROBLEMS TO LOOK FOR: Extensive corrosion except in very dry climates; non-originality, particularly of engine and bumpers; old age and neglect.
CLASSIC CAR STATUS: * * * Set to appreciate.

By the end of 1938, S.S. Jaguar production had reached 5000 a year and was rising fast; so much so that Lyons was already thinking of making his own engine and chassis to reduce his suppliers' profits. Those plans had to be shelved, however, during the early part of the Second World War. Nevertheless, time was found during firewatching sessions to begin designing what was to be the world's first twin-overhead-camshaft hemi-headed mass-production engine.

When peace came, one of Lyons' first moves was to drop the initials S.S. which were now tainted by their association with Hitler's Nazi corps, and to call his cars, simply, Jaguars. Before the war, S.S. Jaguars had sold well abroad, so Lyons was able to obtain enough of Britain's limited steel supplies to launch a new export drive and thereby bring in badly-needed foreign currency.

As work continued on development of the new XK engine, as it was called, the Jaguars produced immediately after 1945 were essentially the same as those built before the war.

◁ ***The S.S. 100 sports car was the most glamorous early Jaguar. Patrick Collection.***

▽ ***The Jaguar XK120 of 1948, the first mass-produced car with a twin-overhead-cam engine.***

The new engine was so revolutionary that it could not be put into mass production immediately and similar problems applied to a new chassis with independent front suspension which Lyons intended as the basis for a world-beating new luxury saloon car. Consequently, the new engine was mounted in a shortened version of the chassis to make a sensational sports car, the XK120, in 1948. The reasoning was that sports cars sold in far smaller volumes than saloon cars, so the XK120 could be built virtually by hand as a valuable publicity medium, while Jaguar waited in a queue for the giant Pressed Steel Company to prepare tooling for the new saloon car's body panels.

However, the result was quite different from that expected by Jaguar. The engine – still in short supply – produced so much power and torque that the XK120 was capable of an easy 120 mph (193 km/h), enabling it to outperform all but pure racing cars. Also, not only did it have one of the most beautiful bodies ever produced by Lyons, but it was priced at only £998 at a time when anything with comparable performance and beauty cost many times more.

The demand for the XK120 was so heavy that Jaguar had no choice but to tool up for mass production of what was originally intended to be only a limited-production machine!

SUMMARY

Plus points

Superb styling.

Exceptional performance.

Safe handling for its time.

Alloy body on early examples.

Durable mechanics.

Very reliable once in good condition.

Many spares readily available.

Minus points

Original drum brakes far inferior to those of later XK150, which had less attractive styling.

Tendency to extensive corrosion in average climates.

C-TYPE

YEARS MADE: 1951–3.
ENGINE: In-line six-cylinder, 3442 cc, twin overhead camshafts, typically 200 bhp at 5800 rpm.
FRAME: Tubular.
PERFORMANCE: Typically 143 mph (230 km/h).
CURRENT PRICE GUIDELINE: **** It will continue appreciating in proportion to the later D-type.
PROBLEMS TO LOOK FOR: Non-originality, especially in engine and brakes (this can extend to completely fake cars), plus deterioration through age and mistreatment.
CLASSIC CAR STATUS: *** Improving as more potential buyers for D-types come forward than sellers.

When the new Mark VII saloon was eventually launched in 1950, it caused as big a sensation as – and even greater demand than – the XK120 sports car. The reasons for its success, especially in an American market pioneered by the XK120, were that it was extremely good-looking, it could carry up to six passengers in great comfort, it could exceed 100 mph (161 km/h), and, above all, it was relatively cheap since Lyons had gambled on its selling in large quantities.

Understandably, the Mark VII proved an exciting competition car, and its sporting career spanned several years. One of its most memorable victories was in the 1956 Monte Carlo Rally.

The Mark VII saloon and the XK120 formed the basis for Jaguar's success in the early 1950s. The XK120 did so well on the racetrack and in rallies that it was developed in 1951 into the C-type for competition. With an uprated XK engine, lightweight tubular chassis (which cost far more to make than the basic car's twin girder design) and more slippery body shape, the C-type promptly won the world's greatest sports car race at Le Mans and, when fitted with revolutionary new disc brakes, went on to win again in 1953.

SUMMARY

Plus points

Excellent performance.

Durable mechanics.

Practical, nostalgic high-performer in numerous historic events.

Minus points

Handling can be on the wild side for the tyro.

Lives forever in the shadow of the D-type.

▽ ***Jaguar's first purpose-built competition car, the C-type, boosted sales immensely by winning twice at Le Mans. Nigel Dawes Collection.***

D-TYPE

YEARS MADE: 1954–7.
ENGINE: In-line six-cylinder, twin overhead camshafts, typically 3442 cc, 250 bhp at 5750 rpm.
FRAME: Monocoque with subframes.
PERFORMANCE: Typically 170 mph (274 km/h).
CURRENT PRICE GUIDELINE: * * * * *
Much-sought-after classic, excellent gilt-edged investment.
PROBLEMS TO LOOK FOR: Non-originality, especially in engine, numerous replica cars – some superb copies – to trap the unwary.
CLASSIC CAR STATUS: * * * * * Will appreciate even beyond this status as one of the greatest and most advanced sports racers of a glamorous era.

Helped by numerous competition victories and a reputation for quality equal to their appearance and perfomance, Jaguar cars sold in phenomenal numbers in the early 1950s. Makers such as Ferrari and Maserati responded with ever-faster cars, but they were met in 1954 by an even more sensational sports racer, the Jaguar D-type.

The most extraordinary feature of the D-type was the bodywork, designed by former Bristol Aircraft Company aerodynamicist, Malcolm Sayer. Not surprisingly, it resembled a plane's fuselage in shape and construction.

Still powered by the XK engine, the D-type achieved three successive victories at Le Mans between 1955 and 1957, to take Jaguar to a new high peak of adulation.

△ ***The Jaguar D-type took its inspiration from aircraft practice. Nigel Dawes Collection.***

SUMMARY

Plus points
- Fantastic appearance and performance.
- Tremendous nostalgic appeal.
- Durable mechanics.
- Numerous events in which it can compete with great distinction.

Minus points
- Handling can be rather wild for the tyro.
- Many very good copies to trap the unwary.
- Widespread non-originality of engines.

JAGUAR MARK II

YEARS MADE: 1959–67.
ENGINE: In-line six-cylinder, twin overhead camshafts, typically 3781 cc, 220 bhp at 5500 rpm.
FRAME: Integral.
PERFORMANCE: Typically 125 mph (201 km/h).
CURRENT PRICE GUIDELINE: * * Starting to appreciate, set for dramatic rise in top models.
PROBLEMS TO LOOK FOR: Extensive corrosion in other than very dry climates; non-originality can be a problem in 3.8-litre overdrive wire-wheeled top model.
CLASSIC CAR STATUS: * * Underrated because it is a saloon car; sweet 3.4-litre variant especially underrated.

The Mark VII saloon was to be developed into Mark VIII and Mark IX forms, and the XK120 into the similar XK140 and XK150, but a new, even more popular Jaguar saloon had been launched in 1955. This was the medium-sized, five-seater Mark I, powered again by variants of the XK engine. As an even faster and more economical saloon than the big Marks VII-IX, it was to become a top-seller throughout the world, particularly when developed into a Mark II form in 1959. When fitted with the ultimate 3.8-litre XK engine, this was the car that became beloved of both businessmen and bank robbers, with a searing performance of up to 125 mph (201 km/h), and it went on to dominate the newly popular production saloon car racing.

△ ***The Mark II was one of Jaguar's most beautiful – and fastest – saloons. Owner: Graig Hinton.***

SUMMARY

Plus points
- Beautiful styling.
- Classic walnut and leather interior.
- Superb performance and comfort.
- Reliable mechanics.
- Good availability of spares.

Minus points
- Tendency towards extensive corrosion in bodyshell.
- Expensive to restore, especially interior.
- Tail-happy handling for the tyro.

E-TYPE

YEARS MADE: 1961–74.
ENGINE: 1961-71 in-line six-cylinder, twin overhead camshafts; 1961–4 3781 cc, 265 bhp at 5500 rpm; 1964–7 4235 cc, 265 bhp at 5500 rpm; 1967–71 typically 200 bhp at 5500 rpm; 1971–4 V12, single overhead camshaft, 272 bhp at 5850 rpm.
FRAME: Monocoque with subframes.
PERFORMANCE: Typically 150 mph (241 km/h).
CURRENT PRICE GUIDELINE: ** Set to appreciate rapidly for cars in good condition.
PROBLEMS TO LOOK FOR: Rampant corrosion in bodyshell in any other than very dry climates, non-originality in early examples.
CLASSIC CAR STATUS: *** Underrated as one of the greatest performers and trendsetters.

Jaguar sales reached a new high when the D-type was effectively converted into production form with the advantage of new independent rear suspension as the E-type in 1961. It had steel body panels and used the 3.8-litre engine that had originally been used in the XK150S.

Never before had such a high-performance sports car been offered at such a low price: it was capable of 150 mph (241 km/h) for £2000. This was one-third of the price of a Ferrari, and into the bargain the E-type was far more reliable and readily available. Predictably, it became the extremely popular car of pop stars and ordinary people alike.

SUMMARY

Plus points

- Fabulous styling and performance.
- Reliable mechanics.
- Great nostalgic appeal for the 1960s.
- Better ride, handling and traction than earlier Jaguars.

Minus points

- Tendency to extensive corrosion in monocoque that is very expensive to repair.
- Brakes on early examples not in keeping with potential performance.
- Bulky styling of two-plus-two variant.
- Overlight steering in V12 examples.

XJ6

YEARS MADE: 1968–86.
ENGINE: In-line six-cylinder, twin overhead camshafts, typically 4235 cc, 245 bhp at 5500 rpm; XJ12 variant from 1972 has V12 engine, single overhead camshaft, typically 272 bhp at 5850 rpm.
FRAME: Integral.
PERFORMANCE: XJ6 typically 127 mph (204 km/h); XJ12 typically 145 mph (233 km/h).
CURRENT PRICE GUIDELINE: * to ** According to year. One to lay down for rapid future escalation.
PROBLEMS TO LOOK FOR: Terminal corrosion, mechanical wear.
CLASSIC CAR STATUS: *** Will rise even beyond this, especially the XJ12.

Aware that his big saloon was beginning to look dated, Lyons introduced a spectacular new large saloon car, the Mark X, in 1962; it was powered by the seemingly ageless XK engine. However, although the Mark X was as well-designed as his earlier cars, it did not achieve their spectacular expansion in sales. This relative failure was due to the fact that fuel was still very cheap in the United States and increasingly large and powerful V8 engines could match the Mark X on acceleration, or even marginally beat it – albeit at the cost of a massive fuel consumption. This situation was contained when the XK engine was enlarged to 4.2 litres to give more pulling power; but already Jaguar was working on a new V12 engine.

The old XK unit was so smooth and durable, however, that development of the V12 took many years before it could offer a great deal more. Meanwhile, Jaguar had been forced to merge with other companies to form British Leyland in 1967: in Jaguar's case the reason behind this move was so that supplies of body panels from Pressed Steel could be maintained.

The formation of British Leyland enabled Jaguar to go ahead with what was to be another world-beating saloon, the XJ6, which replaced both the Mark X's development, the 420G, and all variants of the Mark II saloon in 1968. Demand was so heavy for this extremely sophisticated new car which set new standards of riding comfort, handling and silence, that it was to continue in production with the XK engine only until 1972.

▽ ***The Jaguar E-type – produced in drop-head and fixed-head forms. Owner: Will Athawes.***

△ ***The Jaguar XJ6, in its Mark III form, restored the British firm's reputation for grace, space and pace.***

An immensely smooth and powerful new V12 engine was launched in 1971 – with the XK120's example in mind – in a development of the E-type, before it went into the XJ's bodyshell to create an additional model, the 145 mph (233 km/h) XJ12, in 1972. Sales soared again before the 1973 oil crisis hit this car which, inevitably, used far more fuel than the XJ6. Nevertheless, it was so good that it survived in more limited production as its American rivals fell by the wayside, with an XJ-S two-plus-two sporting version launched in 1975.

These cars, in Mark two and Mark three development forms, carried Jaguar through the crisis years of British Leyland's decline as a major motor manufacturer until the day came when Jaguar could be relaunched as a private company. With demand for the XJ6, XJ12 and XJ-S as strong as before the oil crisis, in 1986 Jaguar was able to launch a new range of XJ6 cars under a dynamic new leader, John Egan. Despite the passage of time and changing tastes, these cars still showed all the style that had been the hallmark of Sir William Lyons.

The new XJ6 became an instant best-seller, especially in the all-important American market. Once the build programme had swung into top gear, Sir John was able to point out triumphantly:

'We have got our quality right and people are rushing to buy our cars, particularly in the United States, which will be taking 25,000 – worth more than £650 million in exports to Britain – in the first full year of production.'

SUMMARY

Plus points

- Excellent performance – magnificent on the XJ12.
- Superb handling and ride.
- Extreme silence and comfort.
- Durable mechanics.
- Good availability of spares.
- Lovely interior on early models.

Minus points

- Tendency to rampant corrosion in bodyshell.
- Very high restoration costs.

JENSEN

The brothers Richard and Allan Jensen earned a reputation for good body styling in the 1930s with the Avon Standard, which competed in the S.S. market, before launching their own car with a Ford V8 engine in 1936. Over the next 40 years, the Jensen marque was to become a byword for performance and technological sophistication.

The first Jensen of 1936 was developed into the PW (for Post War). The original Ford V8 engine was replaced with a 4-litre Austin unit, so helping to establish closer ties with the Austin Motor Company. Jensen's stylist, Eric Neale, then designed a sporting body for Austin to put into production on the small A40 chassis, while Jensen produced a much larger Interceptor grand touring car in 1949. These Interceptors were made only in small quantities, however, because Jensen's main work became body production for the new Austin-Healey 100 (see page 28). Nevertheless, the big, heavy Interceptor continued to be listed until 1958.

CV8

YEARS MADE: 1962–6.
ENGINE: V8, overhead valves, pushrods, 1962–6 5916 cc, 305 bhp at 4800 rpm; 1964–6 6276 cc, 330 bhp at 4600 rpm.
FRAME: Tubular with steel platform.
PERFORMANCE: 130 mph (209 km/h).
CURRENT PRICE GUIDELINE: * * Undervalued in terms of performance, lack of corrosion and comfort.
PROBLEMS TO LOOK FOR: Body and paint damage expensive to repair.
CLASSIC CAR STATUS: * * Set to appreciate.

From 1954 Jensen's bestseller was the 541 – a smaller, much sleeker, version of the Interceptor with a glass-fibre body. In 1956, the 541 became the first British production car to have disc brakes on all four wheels, and in a more radical state of tune, as the 541R, its performance was almost as good as that of the rival Jaguar XK150 (see page 88).

Jensen could never make the 541R in Jaguar-like quantities, however – partly because of its body building contracts which were to include one from Volvo – so the 541 was inevitably rather expensive. Consequently, in 1960 the company decided to make it a good deal more luxurious than a Jaguar, redesigning it around a 4-in (10-cm) wider bodyshell as the 541S, so that more profit could be made from each car. Then problems with the supply of high-performance parts for what was really a truck engine led to a fall-off in performance, and Jensen followed the Bristol path in 1962 by redesigning the 541S around Chrysler's 5.9-litre V8 power unit as the CV8.

This model, with its maximum speed in excess of 130 mph (209 km/h), took Jensen back into the supercar league. The chassis, developed from the 541S, was extraordinarily strong, being formed from two massive steel tubes with a steel platform centre, braced by an equally rigid scuttle. The body, still in glass fibre, was similar to that of the 541S except that the nose was restyled to take a four-headlamp lighting system in a distinctive 'slant-eyed' format that has yet to be matched.

Production of the CV8 continued at the rate of about 200 cars per year, while Jensen carried on making the Austin-Healey 3000 body (see page 30), supplemented by Volvo's P1800 bodyshell and that of the Sunbeam Tiger. Because the CV8 was produced in such small quantities, it could be revised from time to time without causing too much disruption.

Inevitably, a Mark II version followed in 1963 with a larger, 6276 cc Chrysler V8 in 1964 and then a neater Mark III in 1965.

It was in this year that an experimental CV8 was fitted with a four-wheel-drive system developed by the Ferguson tractor firm. This car also pioneered the use of an anti-lock braking system by Dunlop.

SUMMARY

Plus points

- Rarity value.
- Performance.
- Reliability.
- Pulling power of engine.
- Lack of corrosion in glass-fibre bodyshell and massive chassis.
- Comfort.

Minus points

- Strange styling.
- Cost of top-class repairs to bodyshell and interior.

▽ ***The slant-eyed four-headlamp system of the Jensen CV8 gave the model a distinctive appearance.***

FF

YEARS MADE: 1966–71.
ENGINE: V8, 6276 cc, overhead valves, pushrods, 325 bhp at 4600 rpm.
FRAME: Tubular with steel platform.
PERFORMANCE: 133 mph (214 km/h).
CURRENT PRICE GUIDELINE: * * *
Starting to appreciate after years in the doldrums; will eventually become very valuable.
PROBLEMS TO LOOK FOR: Corrosion in body panels, mechanical mistreatment.
CLASSIC CAR STATUS: * * * *
Technological marvel.

In the early 1960s, Jensen saw problems on the horizon: the mooted replacement for the Austin-Healey 3000 was to be simply a re-badged MGC (see page 129) produced at Abingdon, and Sunbeam's Tiger was unlikely to carry on for long with its Ford V8 engine when the parent Rootes Group became part of Chrysler. It therefore became imperative for Jensen to increase the production rate of its own car to make up for the shortfall in outside body work. The idea of a steel body fitted in well with the production capacity, but there was little time in which to style a new body and acquire the special tooling that would be needed for its production. In the event, the Italian firm Touring performed wonders, designing an attractive new body with a distinctive 'goldfish bowl' rear window like that of Maserati's Mistral (see page 115), while Vignale managed to produce the tooling in time for the new car to make its début in 1966.

This sensational Interceptor used substantially the same running gear as the CV8, except that there was also now a longer four-wheel-drive version with anti-lock brakes called the FF (for Ferguson Formula). The Interceptor was an immediate success, and the FF would have been an even greater one had it been easier to build. As it was, the factory could produce only 300 FFs by 1971, at which point the FF had to make way for a new Lotus-engined Jensen-Healey sports car, expected to compete with the Datsun 240Z (see page 56) in the American mass market. However, reliability problems hit sales, and then the 1973 oil crisis made it very hard to sell any sporting car.

Jensen managed to soldier on with the Jensen-Healey and the Interceptor – now up to 7.2 litres, and available in convertible form from 1974 – but in 1976 cash flow problems finally forced the company out of business.

SUMMARY

Plus points

- Excellent styling.
- Very reliable engine, sweeter than later 7.2-litre.
- Massive pulling power and grip.
- First-class handling and braking for 1960s' car.
- Strength of chassis.
- Front-seat comfort.
- Advanced technology.

Minus points

- Complexity of transmission and braking system needs careful maintenance.
- Neglected body panels suffer badly from corrosion.

◁ ***The FF, a longer four-wheel-drive version of the sensational Jensen Interceptor.***

LAGONDA

The first Lagondas were the creation of an American called Wilbur Adams Gunn, who emigrated to England and married a wealthy lady called Constance. He built his motor cycles, and then cars, in a works on her land by the River Thames at Staines, Middlesex, calling the machines Lagondas after the Indian name for his family's home town on the banks of the River Ohio. Athough the Lagonda company was bought by David Brown, owner of Aston Martin, in 1948, the marque still lives on.

LAGONDA M45R

YEARS MADE: 1934–8.
ENGINE: In-line six-cylinder, 4493 cc, overhead valves, pushrods, 141 bhp at 3800 rpm.
FRAME: Twin girders, cross-braced.
PERFORMANCE: 103 mph (166 km/h).
CURRENT PRICE GUIDELINE: * * *
Undervalued compared to rival Bentleys.
PROBLEMS TO LOOK FOR: Non-originality, deterioration through age.
CLASSIC CAR STATUS: * * * Far better than a 1930s' Bentley.

Early Lagondas were extremely well made, establishing the marque with a reputation for very high quality before Gunn died in 1920. The firm was then reorganized along military lines by Brigadier-General Francis Metcalfe to provide handsome 3-litre rivals for W. O. Bentley's similar capacity cars, and it somehow survived the desperate days of the Depression that led to Bentley being taken over by Rolls-Royce. An honourable man, Metcalfe included provisions for Lagonda going into receivership when a syndicate of amateur racing drivers appealed to him – just like W. O. Bentley before – to produce a 4½-litre version of his 3-litre car!

The resultant Lagonda M45, which acquired its name from the fact that it had a 4.5-litre engine produced by the independent manufacturer Meadows,

△ ***The M45R represented the magnificent six-cylinder Lagonda in its ultimate form.***

was no mean performer despite its great bulk, retaining the 3-litre's excellent handling and superb brakes. But the really memorable thing about it was the extraordinary pulling power of its six-cylinder in-line engine that made it very easy to drive. All told it was a far better bargain than the costly new 3½ litre Rolls Bentley, especially when it was uprated to near road-racing specification as the M45R, or Rapide.

Lagonda was dealt a bitter blow when Metcalfe, who had kept it going throughout the Depression, succumbed to cancer. Then the market for big sports cars collapsed for a while in 1935 when driving tests, pedestrian crossings and the 30 mph speed limit were introduced simultaneously. The works at Staines became clogged with unsold Rapides, and a receiver was appointed. Lagonda had powerful friends, however, and none better than solicitor Alan Good and former Bentley boys like Dr J. Dudley Benjafield, who started to form a consortium to keep the firm alive. In fact, Benjafield and the Lagonda specialist garage Fox and Nicholls contrived to enter Rapides at Le Mans in 1935 with outstanding results: the Fox and Nicholls car won against the fastest machinery in the world. Such good fortune would have been the making of many other firms, but the publicity was of little immediate help to Lagonda as it was pronounced bankrupt the following day.

The Le Mans success did serve to convince W. O. Bentley, however, that an approach made to him by Good was worth while. He managed to escape the shackles of Rolls-Royce, who had bought his name and services along with his firm four years earlier, and declared his support for Good on that fateful day. With the win at Le Mans and the prospect of W. O. Bentley as chief designer, Good was able to raise enough money to take over Lagonda.

SUMMARY

Plus points

Engineering on a magnificent scale.

Enormous pulling power.

Massive durability.

Often marvellous coachwork.

Minus points

Restoration costs.

Lives in the shadow of the later V12, which offered nothing less than six extra cylinders.

LAGONDA V12

YEARS MADE: 1938–9.
ENGINE: V12, 4480 cc, single overhead camshaft, 175 bhp at 5500 rpm.
FRAME: Twin girders, cross-braced.
PERFORMANCE: 104 mph (167 km/h).
CURRENT PRICE GUIDELINE: * * * * Over-rated compared to earlier LG45 Rapide, but still a great classic because of its 12 cylinders.
PROBLEMS TO LOOK FOR: Deterioration through age.
CLASSIC CAR STATUS: * * * * Lovely bodywork helps compensate for slightly disappointing engine.

Good's avowed intention in 1935 was to take the notional title of best car in the world from Rolls-Royce. He pursued this objective with a one-model policy; offering too many variants had been the real reason for Lagonda's earlier problems. Bentley, for his part, started designing what he hoped would be that very car.

Initially, the old Meadows engine was uprated into Sanctions 1, 2 and 3 form – so called because its original manufacturer, Harry Meadows, sanctioned the changes – with stylist Frank Feeley producing outstanding bodies. In their ultimate form, these Lagondas were known as LG45s, the 'LG' standing for Lagonda-Good. But all were intended only as interim models on the way to a virtually identical car powered by the new 4.5-litre V12 engine that Bentley was designing. When this car eventually appeared, towards the end of 1938, war was on the way and there was not enough time for the extensive development it needed. However, such is its multi-cylindered glamour that it is the V12 Lagonda which more than any other conjures up romance.

△ ***The V12 was Lagonda's attempt to take the title of 'The Best Car in the World'. Patrick Collection.***

SUMMARY

Plus points

Beautiful bodywork.

Smooth-running engine, even if it does not set the world alight!

Minus points

Expensive compared to six-cylinder models.

High costs of restoration.

LAGONDA SALOON

YEARS MADE: 1976 to date.
ENGINE: V8, 5340 cc, twin overhead camshafts, 397 bhp at 6200 rpm.
FRAME: Steel platform.
PERFORMANCE: 140 mph (225 km/h).
CURRENT PRICE GUIDELINE: ** to **** Depending on age, massive depreciation from high original cost.
PROBLEMS TO LOOK FOR: Mistreatment, deterioration through lack of the detailed maintenance necessary for such a car. Many originally exported to Middle East and suffered badly from sand corrosion.
CLASSIC CAR STATUS: ** Set to appreciate.

During the war, Bentley began designing an entirely new car, which was at last put into production, in simplified form, when Lagonda was bought by David Brown, who already owned Aston Martin. Its engine found its way into the Feeley-styled Aston Martin DB2 (see page 21), while Lagondas continued as more luxurious and refined versions of the contemporary Aston Martin touring cars.

Early examples produced in the 1950s followed similar body lines to the pre-war V12, until they were replaced in 1962 by a far more sporting version of the Aston DB4 (see page 22), called the Rapide. This was followed briefly by a stretched, four-door version of the DBS V8 (see page 23) before the freelance stylist William Towns was commissioned to produce what would be one of the most sensational sights of the 1970s in the wedge-shaped Aston Martin Lagonda four-door saloon.

Despite the fact that Towns and Aston Martin had only eight months in which to design and build the prototype, they came up with a real winner. This low-slung luxury saloon is powered by the current Aston Martin 5.3-litre V8 engine and is still in production at the time of writing.

SUMMARY

Plus points

- Sensational styling for the 1970s.
- Superb quality of materials used in construction.
- Muscular engine.
- Excellent Chrysler automatic transmission.

Minus points

- Poor reliability if not maintained to exacting standards.
- Cramped cabin for what is supposed to be a big saloon car.
- Very heavy bodyshell deadens handling.
- Very high costs of restoration to original specification.

▽ ***The Lagonda saloon, styled by Towns and produced by Aston Martin, offered one of the most stunning bodies on a British car.***

LAMBORGHINI

The fabulous Italian marque Lamborghini was born out of an inevitable holocaust at a time when Ferrari and two or three others were dominating the exotic car market with machinery which was far from perfect technically and often left much to be desired.

The people who bought such cars wanted something exclusive that could outperform anything else they were likely to meet and outclass anything else technically . . . or at least appear to do so.

They were also willing to pay handsomely for such a privilege, so the profit margins in this field were far higher than normal.

The designers, on the other hand, were often frustrated, for they could not achieve the fame and fortune they felt they deserved, because manufacturers such as Ferrari were inclined to be extremely conservative, not to say reactionary.

The explosion came when six of Ferrari's top designers and technicians walked out in 1961, taking with them not only their talents, but also many of the ideas they would have liked to have seen put into practice on Ferrari road and racing cars.

These key figures included Carlo Chiti, who found his way back to Alfa Romeo (see page 14), and Giotto Bizzarrini, who became involved with Iso, among other companies (see page 82).

Dissatisfied Ferrari customers included tractor and central-heating tycoon Ferruccio Lamborghini, a former Fiat Topolino tuner (see page 72), who was sure he could produce a better car than Ferrari. He was not interested in making a racing car: other manufacturers had shown that they could beat Ferrari on the track, so there was nothing special in that, and the glory would fade when they were defeated.

But he reasoned that nobody had built a road car which was definitely better than a Ferrari, despite the latter's faults; and the glory of building the best road car in the world would last forever!

LAMBORGHINI 350GT

YEARS MADE: 1963–5.
ENGINE: V12, 3497 cc, twin overhead camshafts, 360 bhp at 8000 rpm.
FRAME: Steel platform.
PERFORMANCE: 168 mph (270 km/h).
CURRENT PRICE GUIDELINE: *** May appreciate despite controversial styling.
PROBLEMS TO LOOK FOR: Mistreatment, non-originality.
CLASSIC CAR STATUS: **** First of a historic marque, performs better than it looks.

When Ferruccio Lamborghini set out to produce his first car, he was tempted by the idea of a 16-cylinder engine to demonstrate irrevocably that it was superior to Ferrari's V12, but he realized that such an engine would be impractical for a road machine. So he hired Bizzarrini to design a Lamborghini hemi-head engine with twin overhead camshafts on each bank of cylinders: this would make it superior to Ferrari's contemporary single cam units. Initially this V12 had a 3.5-litre capacity as that was sufficient to put it ahead of the 4-litre Lampredi engine which would later be used in Ferrari's 330GT. Once he had designed this engine, which was his statement of what he thought a Ferrari Grand Prix engine should be like, Bizzarrini left to concentrate on other projects, notably racing cars.

Lamborghini then hired a bright young team led by Giampaolo Dallara, Paolo Stanzani and New Zealander Bob Wallace, to develop his traditional front-engined grand tourer. Like Bizzarrini, they really wanted to produce a racing car, but they fell in with Lamborghini's wishes because they realized that he was giving them an opportunity to demonstrate their talents. So far as the chassis and running gear of the new car were concerned, these amounted to a roadgoing version of what Bizzarrini would have liked to have seen on the Ferrari 250GTO (see page 68). The suspension, in particular, was far superior in that it used a sophisticated independent system at the back, rather than the old-fashioned live axle and leaf springs.

Lamborghini also had strong ideas on body styling which, unfortunately, resulted in some compromises before his new 350GT was unveiled in 1963. He demanded a nose like the Jaguar E-type's and a tail like that on Aston Martin's DB4, plus many of the features seen in Ferrari Superfasts and the Chevrolet Corvette Stingray (see pages 89, 22, 65 and 44). Stylist Franco Scaglione did his best, but the body was far from a classic. Happily, however, Touring had managed to smooth out a lot of the clashing lines by the time it put the two-seater bodywork into production.

Sales were slow at first because of a financial crisis in Italy, although numerous potential customers were excited by the mechanical specification which gave a much higher performance than rival Ferraris, Maseratis and Aston Martins.

To improve the situation, within months Dallara was busy turning the 350GT into a more saleable two-plus-

▽ Lamborghini's 350GT offered a sensational performance in advance of its rivals.

△ ***The Lamborghini Miura took the 1966 Geneva Show by storm. Owner: Michael Gertner.***

two seater. The new model used steel panels rather than aluminium for quantity-production economy, but they weighed more, and the engine capacity had to be increased to 4 litres to maintain the performance advantage. Introduction of the new car, called the 400GT 2+2, was effectively delayed until 1966, though, so that a surplus of old body parts could be used up; some of these were used to update the appearance of the earlier 350GTs. When fitted with 4-litre engines, these rejuvenated 350GTs were called 400GTs, while one-off variants used special bodies on 3.5-litre or 4-litre chassis. This line was then extended in 1967 with the Islero, a two-plus-two with a simpler body design by Ferruccio Lamborghini himself.

SUMMARY

Plus points

- Excellent performance and handling.
- Marvellous engine, even sweeter than later 4-litre.
- Goes better than a contemporary Ferrari grand tourer.
- Reliable mechanics if treated properly.

Minus points

- Sometimes poor build quality of bodywork.
- Tendency towards corrosion of body panels and supports.
- Controversial styling.
- Needs racing car-style maintenance.
- Very high cost of spares.

MIURA SV

YEARS MADE: 1971–3.
ENGINE: V12, 3929 cc, twin overhead camshafts, 385 bhp at 7850 rpm.
FRAME: Monocoque and subframes.
PERFORMANCE: 170 mph (274 km/h).
CURRENT PRICE GUIDELINE: * * *
Appreciating fast, earlier Miura and Miura S models cheaper.
PROBLEMS TO LOOK FOR:
Mistreatment, lack of vital detailed maintenance.
CLASSIC CAR STATUS: * * * * Underrated against Ferrari opposition as best of a new breed of mid-engined supercars.

Customers for the GT range were disappointingly few, so stylist Dallara pressed Lamborghini to initiate a racing programme in the hope of boosting sales. As part of his campaign, he had made a detailed study of the most exciting sports racing car of the day, the Ford GT40 (see page 75). Lamborghini listened, but did nothing. He considered racing a waste of money and blamed the GT cars' bodywork for their slow sales.

While Dallara, Stanzani and Wallace spent their spare time working on a prototype mid-engined GT40-style car, Lamborghini talked to Nuccio Bertone, the stylist whose great rival, Pininfarina, was doing a lot of business with Ferrari at the time. When, in 1965, the young designers showed Lamborghini their new chassis with the V12 engine set transversely, Ferruccio said that he had no intention of making a racing car: instead, he would make their prototype into something better – an immortal GT car with a body by Bertone, who just happened to have a brilliant new stylist in Giorgetto Giugiaro. The task was given to Giugiaro, but he was about to leave to work for Ghia, so he passed it on to his assistant, Marcello Gandini. Here was another young man anxious to prove himself, and the result of his efforts effectively took the 1966 Geneva Show by storm.

The first P (for Posteriore, denoting that the engine was behind the driver) 400 Miura was so attractive that orders flowed in despite its very high price. Lamborghini was overjoyed, saying: 'This means so much more than winning a race which everybody will forget within a few weeks. People will remember my name for years. . . .'

The Miura did not rest on its laurels, however. In 1969 the Miura S appeared, and in 1971 came the ultimate version – the SV (Super Veloce, or 'super fast'). The Miura SV boasted revised suspension and increased power among its new features.

SUMMARY

Plus points

- Stunning performance.
- Handling much improved over earlier under-tyred Miura and Miura S.
- Far more reliable power train than earlier mid-engined models.
- Excellent styling.
- Supreme historic supercar.

Minus points

- Needs racing car-style maintenance.
- Poor build quality of bodywork and interior.
- High restoration costs.
- Very high cost of spares.

△ ***The long, low, sensationally fast Espada was Lamborghini's idea of what a Rolls-Royce should be.***

ESPADA III

YEARS MADE: 1973–8.
ENGINE: V12, twin overhead camshafts, 345 bhp at 7800 rpm.
FRAME: Steel platform.
PERFORMANCE: 155 mph (249 km/h).
CURRENT PRICE GUIDELINE: ** to ***
Marvellous value for money when in good condition; will appreciate.
PROBLEMS TO LOOK FOR:
Mistreatment, lack of vital detailed maintenance, deterioration of very expensive bodywork.
CLASSIC CAR STATUS: ** Grossly underrated, unrivalled as the only four-seater with real performance and exotic style.

Soon after the Miura's introduction, plans were progressing for what Lamborghini called an Italian Rolls-Royce; a full four-seater saloon using the same basic mechanical components to give it an extraordinary performance. Lamborghini's main dictate – apart from the fact that it had to have four comfortable seats – was that the new car (to be called the Espada) should be long and low, not 'upright and stodgy' like a Rolls-Royce. Gandini, who was riding on the crest of a wave of adulation after his work on the Miura, approached his brief with enthusiasm. So that the passengers would not feel claustrophobic, he had the doors of the prototype, which opened upwards like a gull's wings from low on the body, made entirely of glass. Lamborghini was horrified! He promptly rejected the concept, saying that he did not want everybody to ogle his lady's legs. . . . Consequently, the final design for the Espada saloon car was switched to that of a Pirana show model which Bertone had built on a Jaguar E-type chassis (see page 89).

When it went into production in 1968, this long, low and very wide car yielded a maximum speed of 150 mph (241 km/h) making it the world's fastest four-seater saloon. Predictably, the Espada caused a sensation. As with the Miura, orders flowed in.

At the 1972 Turin Show an even finer version of this superb car was revealed to public scrutiny – the Espada III. The model's top speed had now risen to 155 mph (249 km/h) and detail improvements had been made to the body, brakes and suspension.

The Islero was beginning to look very dated by this time, so it was replaced in 1970 by a short-wheelbase two-plus-two version of the Espada, called the Jarama. Sales struggled against Maserati's Ghibli and Ferrari's Daytona, though, because the Jarama was rather heavy, having been based on the much more solid Espada, and as a result could reach only 162 mph (261 km/h) (see pages 114 and 70). Early problems caused by the transversely mounted engine having to share the same oil supply as the transmission – as in Britain's mundane Mini (see page 130) – were cured by complicated new castings, and, among other improvements, the Jarama was fitted with Pirelli's new low-profile tyres to give much better handling.

SUMMARY

Plus points

- Fantastic performance and handling.
- Marvellous engine.
- Full four-seater accommodation.
- Far better steering than earlier Espadas.

Minus points

- Needs racing car-style maintenance.
- Poor build quality of bodywork and interior.
- Very high restoration costs.
- Very high cost of spares.

P300 URRACO

YEARS MADE: 1974–6
ENGINE: V8, 2996 cc, twin overhead camshafts, 265 bhp at 7800 rpm.
FRAME: Integral.
PERFORMANCE: 160 mph (257 km/h).
CURRENT PRICE GUIDELINE: * * *
Grossly underrated as one of the best-handling small GT cars ever made. Easily outperforms contemporary Ferrari and Porsche opposition.
PROBLEMS TO LOOK FOR:
Mistreatment, lack of vital detailed maintenance.
CLASSIC CAR STATUS: * * * Underrated.

Delays in getting the Countach (see page 100) into production were initially due to the fact that Lamborghini was too occupied with trying to expand into Porsche's market for small, mass-produced GT cars. Lamborghini, who habitually drove a Fiat 500 when not using his Islero (the car he had designed himself), saw this venture as an initial exercise before he expanded into mass manufacture.

Porsche's 911 had two small rear seats, and when Ferrari produced the Dino with only two seats (see page 69), Lamborghini became even more determined that his 'mini-Miura' should be a two-plus-two. He was also adamant that it should be produced as far as possible by automation – which meant heavy investment in machine tools – because Italy was suffering badly from industrial strikes.

Porsche had placed its engine behind the rear wheels to liberate enough space for rear seats, leading to tail-happy handling which could catch out the tyro. Stanzani would have none of this, and decided to build a new V8 engine as it would be shorter and take less room than a straight six-cylinder from one half of the existing V12. He opted against the even more compact V6 cylinder configuration because at that time it was hoped that the 2.5-litre V8 could be split down to a 1250 cc in-line four to power a mass-production car like the British Mini.

The resultant Urraco (or 'little bull') needed a great deal of capital investment and development, so it was not possible to get it into production until 1973, by which time the world oil crisis was looming. Car sales everywhere suffered – especially of high-performance models – and Ferruccio Lamborghini, tiring of the problems associated with becoming a mass manufacturer, sold out and retired to a vineyard where he still produces his own wine.

Nevertheless, when the Urraco was sufficiently developed in the form of the P300, with a more powerful 3-litre engine to complement its superb handling and braking, it became one of the best small GT cars ever produced. To date, the Urraco has evolved through the rebodied Silouette into the Jalpa.

△ ***Lamborghini achieved wonders by making the 'mini-Miura' – the Urraco – a two-plus-two.***

SUMMARY

Plus points
- Superb performance.
- Fantastic, sensuous handling.
- Excellent styling.
- Far superior package to earlier single-cam P250 Urraco.

Minus points
- Needs racing car-style maintenance.
- Poorer build quality than a Porsche.

COUNTACH

YEARS MADE: 1974 to date.
ENGINE: V12, twin overhead camshafts, 1974–82 3929 cc, 375 bhp at 8000 rpm; 1982–6 4754 cc, 400 bhp at 7000 rpm.
FRAME: Tubular.
PERFORMANCE: 175 mph (282 km/h).
CURRENT PRICE GUIDELINE: *** to **** depending on age.
PROBLEMS TO LOOK FOR: Mistreatment, lack of maintenance.
CLASSIC CAR STATUS: ***** For years the unrivalled supercar.

As the Jarama was going into production in 1971, an even more stunning car appeared at the Geneva Show. It was a combination of Stanzani's ideas on how the Miura could be radically improved, and Bertone's development of a theme which had originated in his Carabo show car, built on an Alfa Romeo Tipo 33 racing car base, and continued with the futuristic Lancia Stratos of 1970 (see page 104). Problems with the Miura's transverse engine mounting were solved by turning the engine round so that it lay north-south rather than east-west, with the gearbox at the front so that its weight eliminated the Miura's tendency for the front wheels to lift at high speeds.

Gandini's new body was extraordinary: it was the ultimate expression of the wedge shape which had become popular in racing for aerodynamic reasons in that it promoted stability when cornering. The doors opened up like the wings of a beetle (or Carabo) in the most striking fashion. As this car was very much a Bertone project, it received a Bertone name – Countach (pronounced 'coon-tash'), a Piedmontese expression of rude wonder.

It took three years for Lamborghini to get the Countach into production, but it was worth the wait, for the Countach remains the world's greatest supercar. The most recent version, the Countach S, has latest-specification tyres among its improvements.

SUMMARY

Plus points
- Fantastic styling.
- Stunning performance.
- Marvellous engine.
- Superb ride and handling.
- Tremendous strength.
- Low depreciation of new models by supercar standards.

Minus points
- Needs detailed maintenance for self-preservation.
- High cost of spares.

▽ The Lamborghini Countach – pictured in its original form – became the definitive supercar.

LANCIA

The Italian marque Lancia has always had a reputation for producing cars of advanced specification since it was founded by F.I.A.T. racing driver Vincenzo Lancia in 1906. His early cars were excellent, but his most famous, the Lambda, did not appear until the early 1920s.

LAMBDA

YEARS MADE: 1923–31.
ENGINE: V4, single overhead camshaft, typically 2570 cc, 69 bhp at 3500 rpm.
FRAME: Integral.
PERFORMANCE: Typically 75 mph (121 km/h).
CURRENT PRICE GUIDELINE: *** Technological pioneer, deserves higher rating.
PROBLEMS TO LOOK FOR: Plain old age.
CLASSIC CAR STATUS: **** Again deserves higher.

The Lambda's specification was extraordinary at a time when even the greatest marques were still offering what were basically Edwardian designs with lengthy and inefficient in-line side-valve engines, ladder frame chassis which whipped badly, and leaf spring suspension that had hardly progressed since the days of the horse and cart. In stark contrast, the Lambda showed highly original thinking in that it had a space-saving and very efficient overhead-camshaft V4-cylinder engine, and a very stiff all-steel body-cum-chassis with independent front suspension, which meant that it held the road far better than its contemporaries.

Like many other great designers, Vincenzo Lancia did not do the detail work – he left that to a team of specialists – but he was responsible for the concept of the cars he produced. It is said that his vision of the Lambda, the car which would change the face of the motor industry, came one day when he was sailing on Lake Como. As a sudden squall lashed the hull of his boat, he reflected how much stronger it was than the body of a motor car.

For five years he pondered on the problems of making a car as strong as a

ship, realizing that a car using a really rigid chassis would be well-nigh undrivable if it had to use the stiffly sprung beam axle suspension which was in vogue at the time. The front wheels, especially, would be almost constantly bouncing from one pothole to the next, to the detriment of the steering, which relied on keeping the tyres in contact with the ground. Such problems had been overcome on motor cycles, which had to have rigid frames for security rather than a whippy chassis to compensate for a lack of springing. A young draughtsman, Battisto Falchetto, sketched 14 systems of independent front suspension, from which Lancia selected the sliding pillar principle used by Morgan, the English firm which built motor cycle-based three-wheelers (see page 131).

His ponderings on ships' hulls had also brought home to him the advantages of steam engines, in which the cylinders were arranged in a 'V' formation to make the best use of the space at the bottom of the hull and to concentrate their weight low down for maximum stability. It then occurred to Lancia that a V4 engine would be so short that the clutch and gearbox could also be accommodated under a conventional bonnet, so freeing more space for passengers inside the body.

In those days, the occupants of a car normally perched on top of a floorline dictated by the twin girders which formed the chassis, since there was not enough room to sit two abreast inside the chassis rails of a car if it was to be narrow enough to use average-width roads. Again, as he thought about the strength of a ship's hull, Lancia realized that the pointed section at the bottom was the strongest part – and when he saw fishing boats upturned for painting on the shores of Lake Como, he knew he had the answer to his problem. He visualized a transmission tunnel, like that which covered a ship's propellor shaft, running through the centre of his car. Used in conjunction with rigid steel body sides and cross-bracing wherever possible, such as at the radiator, seat backs and tail, this tunnel would be strong enough for him to be able to dispense with the chassis side rails. This would mean that the occupants could sit lower and the car would hold the road better.

The Lambda went through nine distinct series with ever-improving mechanical specifications and becoming increasingly large, roomy and imposing, with open or closed bodywork, until 1931, when it was replaced by a small Artena, a medium-sized Astura and a large Dilambda, all of which used V-shaped engines and independent front suspension.

△ ***The Lancia Lambda was a superbly practical pioneer car with outstanding performance.***

SUMMARY

Plus points
Superbly practical pioneer with extraordinary performance for its age.

Numerous events in which it can compete.

Minus points
Not as glamorous in its appearance as its technical specification and ability warrant.

△ ***Aerodynamics played a large part in determining the Lancia Aprilia's bodywork.***

APRILIA

YEARS MADE: 1937–49.
ENGINE: V4, typically 1486 cc, single overhead camshaft, 48 bhp at 4300 rpm.
FRAME: Integral.
PERFORMANCE: Typically 79 mph (126 km/h).
CURRENT PRICE GUIDELINE: * * Rare outside Italy.
PROBLEMS TO LOOK FOR: Corrosion, old age.
CLASSIC CAR STATUS: * * * Technological pioneer.

While his engineers were busy perfecting the Artena, Astura and Dilambda, Lancia's fertile imagination was giving birth to yet another revolutionary car. He was concentrating not on specific areas, such as handling or fuel consumption, but on how a completely new car could be better in every way than the ones before. Aerodynamics played a large part in his ideas, with nautical principles as a guide – he saw the waves of Lake Como as being very similar to the air streams which a car's body has to part.

The result was Lancia's next great car, the lovely little Aprilia saloon with a beautifully streamlined body shaped like a boat turned upside down, which he patented in April 1934. Once more Falchetto was on hand to make it all work with the unitary construction pioneered on the Lambda. This car broke new ground in that it had torsion bar independent rear suspension instead of a live axle and leaf springs, and was exceptionally economical with fuel due to its efficient streamlining. Technical sensations included the use of hydraulic brakes which were much more efficient than the old cable-operated variety, and windcheating concealed door hinges. The Aprilia became a second memorial to Vincenzo Lancia when it went into production in 1937, the year of his death.

SUMMARY

Plus points

Marvellous practical performer once in good condition.

Years ahead of its time.

Minus points

Decay through age.

No synchromesh for the uninitiated.

AURELIA B20GT

YEARS MADE: 1953–5.
ENGINE: V6, 2451 cc, single overhead camshaft, 118 bhp at 5000 rpm.
FRAME: Integral.
PERFORMANCE: 111 mph (178 km/h).
CURRENT PRICE GUIDELINE: *** Set to appreciate, underrated against Spyder equivalent.
PROBLEMS TO LOOK FOR: Non-originality, corrosion, old age.
CLASSIC CAR STATUS: **** High appreciation in mainland Europe, now extending to other countries.

△ ***Lightweight fastback coupé versions of the Lancia Aurelia were developed into the B20GT rally car.***

When Lancia died, his influence on design was taken over by Vittorio Jano, the great engineer who had been pensioned off by Alfa Romeo as being too old! It was Jano, and his brilliant young assistant, Francesco de Virgilio, who created the third great Lancia, the Aurelia. It was one of the first cars in the world to use radial-ply tyres, but above all it should be remembered as the world's first production car with a V6 cylinder engine and, in B20 form, as the world's first GT car.

The advanced specification of the Aurelia was dictated by the desire of Vincenzo's son, Gianni, to follow his father's innovatory policy. Instead of merely updating the Aprilia, he decided as early as 1943 that it should be replaced by a new car that would be even more comfortable, without sacrificing any of the Aprilia's qualities. De Virgilio achieved wonders in balancing the V6-cylinder unit (needed for extra power), a task that had defeated Lancia as early as 1924. In cooperation with Pinin Farina, Jano retained the Aprilia's pillarless body design, as it offered the easiest access for driver and passengers. Inboard back brakes and a combined rear axle and gearbox called a transaxle, which had been used in Alfa Romeo's 1938 Grand Prix car (see page 10) – Jano's last design for Alfa Romeo – ensured beautifully balanced handling, excellent traction over still-rough roads, and an interior as unimpeded as that of the Lambda. This last feature meant that the saloon version of the Aurelia was able to carry six people in a medium-sized bodyshell, a fact which caused yet another sensation on the model's introduction in 1950.

Lighter fastback coupé versions of the Aurelia did so well in competition that an ultimate low-line version with a 2.5-litre engine, called the B20GT, was developed. This model lived up to expectations by winning the Monte Carlo Rally in 1954.

Subsequently, de Dion rear suspension was transferred from sports racing versions to the saloons, improving the handling even further.

Twenty variants of the Aurelia were made, ranging from long-wheelbase taxis to wooden-bodied estates, while rolling chassis were marketed for specialist coachbuilders, along with a scaled-down version called the Appia. But after 1955, when Lancia was taken over by cement king Carlo Pesenti, the Aurelias became softer and more luxurious, until they were eventually replaced in 1956 by the more weighty Flaminia.

SUMMARY

Plus points

- Excellent high performer with good handling.
- Numerous events in which it can compete.
- Cheaper and more practical than a contemporary Ferrari or Maserati.

Minus points

- Sheer age leads to corrosion.
- Mechanical problems.
- Needs specialized knowledge for maintenance.

FULVIA HF VARIANTE 1016

YEARS MADE: 1969–70.
ENGINE: V4, typically 1584 cc, single overhead camshaft, 132 bhp at 6600 rpm.
FRAME: Integral.
PERFORMANCE: 118 mph (190 km/h).
CURRENT PRICE GUIDELINE: ** Set to appreciate rapidly with new historic rally series.
PROBLEMS TO LOOK FOR: Terminal corrosion, non-originality, mechanical mistreatment.
CLASSIC CAR STATUS: *** Set to appreciate.

The first Lancia revolution of the 1960s came in the shape of the 1961 Flavia, designed by Antonio Fessia of Fiat Topolino fame (see page 72). The Flavia was the first front-wheel-drive car to be built in Italy, with a flat four-cylinder engine of true 'boxer' configuration. Initially it filled a huge gap in the model line-up between the small Appia and the large, expensive Flaminia. In 2-litre form it became Lancia's flagship until 1975, after financial problems led to the company being taken over by Fiat.

Four years later came the Appia's replacement, which was to prove the second great car of the 1960s. Designed by Fessia and with a V4 engine, this new small car was named the Fulvia. The elegant little coupé versions, in particular, were best-sellers for Lancia and became leaders in world rallying alongside the Alpine-Renault A110 (see page 18). For rallying, the Fulvia was first developed as the HF (for High Fidelity) coupé in 1966, and then, with

△ ***Lancia became a leader in world rallying with the Fulvia HF coupé, fitted with a 132 bhp engine.***

a larger 1584 cc engine, as the 1600 HF in 1968. These lightweight cars, using the Flavia's five-speed gearbox, were meant to be used only in competition, but were so attractive that they were produced until 1976.

SUMMARY

Plus points

Agile high-performer, especially on tortuous mountain roads much favoured for historic rallies.

Economical to run.

Minus points

Tendency to terminal corrosion.

Mechanical parts can be overstressed by high standards of tuning.

LANCIA STRATOS

YEARS MADE: 1973–4.
ENGINE: V6, 2418 cc, twin overhead camshafts, typically 190 bhp at 7400 rpm.
FRAME: Monocoque with subframes.
PERFORMANCE: 140 mph (225 km/h).
CURRENT PRICE GUIDELINE: * * * Appreciating rapidly.
PROBLEMS TO LOOK FOR: Terminal corrosion in monocoque, mechanical mistreatment.
CLASSIC CAR STATUS: * * * * Will move higher.

The ever-increasing demands of world rallying came to a head when Ford started fitting what were basically Formula 2 racing engines in the small Escort bodyshell. Then, in 1970, Lancia's attention was taken by Bertone's Stratos showcar, which used a mid-mounted Fulvia 1600 engine and had a rear-hinged opening passenger compartment lid. Like Lamborghini, Lancia saw great potential in this design, but purely for competition use rather than as a supercar built exclusively for the road. In the event, the Lancia Stratos became the world's first rallying supercar, establishing a trend that was to continue to excite race fans everywhere with the thrilling Group B competition of the 1980s.

Lancia completely reworked the Stratos around an immensely strong central monocoque, using the transverse engine and transmission from a Ferrari 246 Dino to give it savage power (see page 69). This arrangement was possible because both Lancia and Ferrari had been controlled by Fiat since 1970. Prototypes of the Stratos were so successful that Lancia decided to homologate it as a special GT car by putting it into very limited production in 1974. It was then developed for all manner of different competitions with various suspension systems, normal or turbocharged engines, two valves or four valves per cylinder head, and a variety of aerodynamic wings until it was replaced in 1978 with the Fiat 131 Abarth (see page 8), which bore a closer resemblance to a standard production car in appearance than the previous version had.

In 1972, the Stratos was followed by an advanced new front-wheel-drive saloon, the Beta, powered by a variety of Fiat engines, and the range was expanded to include spyders and fixed-head coupés in 1974. The introduction of the HPE (for High-Performance Estate) on a Beta base in 1975, was followed by a rapid Pininfarina-styled mid-engined coupé, the Beta Monte Carlo, in 1976. Also new on the scene that year were the Gamma – which was a replacement for Fiat's only large luxury car, the 130 – and the Delta, which was a close relative of the front-wheel-drive transverse-engined five-speed Fiat Ritmo.

The launch of the Audi Quattro ended the Abarth 131's domination of world rallying (see page 25), leading Fiat to develop the distinctive mid-engined Monte Carlo, which was doing well with its Ferrari power on the race track. Yet another supercar, the Lancia Rallye, was brought out for new Group B regulations in 1982. In supercharged form, its 16-valve 1995 cc twin-overhead-camshaft engine put the Rallye, which was still on two-wheel drive, on a par with Audi's sensational Quattro.

Then, when Peugeot went four-wheel drive with its 205 turbo 16 purpose-built rally car, Lancia followed suit in 1985 with the even more sensational Lancia Delta S4. This was both supercharged and turbocharged, to overcome the inherent difficulties with throttle response from turbocharged power units, and these improvements took Lancia once more to the top in road car technology.

SUMMARY

Plus points

Fantastic performance in the hands of a skilled driver.

Savage appeal of macho rally-car image.

Minus points

Tendency towards terminal corrosion in monocoque.

Mechanical parts need racing car-style maintenance.

High restoration costs.

Very twitchy handling can catch out the unwary.

▽ Lancia's wedge-shaped Stratos became the first supercar in world rallying. Owners: Ian Fraser and Nick Mason.

LINCOLN

Lincoln cars have always enjoyed a fine reputation in their native United States for quality of engineering and construction. However, this reputation has not generally spread to every other part of the world because the marque's cars have invariably been produced only in small quantities – by American standards – and rarely exported, because of strong competition in their high price bracket.

The first Lincolns were the creation of Henry Martyn Leland, a toolmaker in the Springfield arsenal during the American Civil War, who went on to manufacture gears and engines. In 1901 he became involved in the car industry, supplying Oldsmobile power plants, and soon began an association with Henry Ford, who was building racing cars nearby. Leland was called in to capitalize on Ford's designs for the road and the embryonic Cadillac Automobile Company was formed in 1902. Then Ford left to start his own company. . . .

By 1909, the Cadillac's accuracy and precision had led to a prosperous business which tempted W. C. Durant to make it part of his emerging General Motors empire. Despite his age, Leland stayed on, introducing the first classic V8-cylinder engine to appear in an American automobile. By 1917, however, Leland had left to set up a company manufacturing the Liberty aero-engine. When the First World War ended, Leland changed to car manufacture. As a staunch Republican, the 75-year-old Leland named the company after the martyred president of his youth: Abraham Lincoln. (It is ironic that when John F. Kennedy was assassinated in 1963, he was riding in the top car of the day – a Lincoln Continental.)

LINCOLN ZEPHYR

YEARS MADE: 1936–9.
ENGINE: V12, 4378 cc, side-valve, 110 bhp at 3400 rpm.
FRAME: Integral.
PERFORMANCE: 90 mph (145 km/h).
CURRENT PRICE GUIDELINE: * * * Rare outside United States.
PROBLEMS TO LOOK FOR: Corrosion, engine trouble, deterioration through age.
CLASSIC CAR STATUS: * * * Pioneer styling, but tends to be overshadowed by the later Continental development.

The first Lincoln cars were magnificent vehicles, but the company soon ran into trouble over alleged – and exaggerated – wartime tax arrears. The Lelands appealed to Henry Ford for help and he bought their company, but they were

▽ Lincoln sales soared when it entered a slightly lower-priced market with the V12 Zephyr saloon.

△ ***Lincoln's second great Continental appeared in 1961 – the ultimate American limousine.***

not comfortable with his more modern methods of car production and left in 1922. Ford's son Edsel then took over, continuing the quality-car theme, but using more modern styling. By the late 1920s, the fast-accelerating Lincoln V8s were beloved of police and gangsters alike. American presidents also bought them, but sales began to decline during the Depression years, coinciding with Lincoln's move to a V12, like its chief competitor, Cadillac.

Sales soared back up when Lincoln entered a less exclusive and slightly lower-priced market with the Lincoln Zephyr of 1936. The V12 theme was followed through in this saloon which was made available in two- and four-door forms. However, the most significant feature was the streamlined styling which had obviously been dictated by aerodynamic considerations.

SUMMARY

Plus points

Attractive styling breakthrough.

Durable mechanical components with exception of engine.

Minus points

Corrosion in bodywork.

Less than reliable engine.

LINCOLN CONTINENTAL

YEARS MADE: 1961–9.
ENGINE: V8, overhead valves, pushrods, typically 7571 cc, 340 bhp at 4600 rpm.
FRAME: Integral.
PERFORMANCE: 123 mph (198 km/h).
CURRENT PRICE GUIDELINE: ** Set to appreciate as one of the better-looking American luxury liners.
PROBLEMS TO LOOK FOR: Corrosion in damp-climate cars, high mileage wear and tear.
CLASSIC CAR STATUS: ** Underrated as the last four-door convertible.

The Zephyr's sleek lines became the basis for stylist Eugene Turrenne Gregorie's much-acclaimed Lincoln Continental coupé in 1939. Spurred on by Cadillac's aggressive new tailfins in 1948, Lincoln replaced the Continental with the slab-sided Cosmopolitan in 1949, complete with a new V8 engine. This was good enough to make the big hardtops, and 'baby' Lincolns based on Ford's medium-priced Mercury range, front-runners in the Carrera Panamericana road race of the early 1950s. Lincoln then pioneered the use of full-colour photography, rather than artists' drawings, in their publicity material from 1954.

The range was revitalized with a Continental Mark II in 1956 and, a year later, this model pioneered the use of dual headlights, which had been illegal in some American states until then. A 'longer, lower, wider look' – which, in reality, meant much larger – for 1958 was not well received, however, and sales flagged far behind Cadillac.

The late 1950s saw Lincoln in the doldrums but, happily, Ford staff stylist Elwood P. Engel had been privately working on ideas for Ford's Thunderbird, which had progressed beyond its original intention of being a Chevrolet Corvette competitor (see page 44).

His simple, unadorned, slab-sided approach was extended just enough to produce Lincoln's next great Continental in 1961.

Because it had a shorter wheelbase than the previous gargantuan '58 model, unique rear-hinged back doors were used for easy access to the back seat.

When this theme was carried over to a drop-head version, it produced the last remaining four-door convertible in the world, staying in production until 1970.

SUMMARY

Plus points

Trendsetting body shape.

Unique late rear-hinged back doors give superb access to rear.

Durable mechanics.

Minus points

High restoration costs aligned to sheer size of vehicle.

All-round drum brakes until 1966.

LOTUS

The most significant car designer of modern times was undoubtedly Colin Chapman, of Lotus. The fact that Chapman's business acumen and imagination equalled his engineering skills ensured the survival of Lotus as one of the great classic marques.

LOTUS MARK 6

YEARS MADE: 1953–6.
ENGINE: In-line four-cylinder, typically Ford 1172 cc, side-valve, 36 bhp at 4400 rpm.
FRAME: Tubular.
PERFORMANCE: Typically 75 mph (121 km/h).
CURRENT PRICE GUIDELINE: * * Set to appreciate gradually.
PROBLEMS TO LOOK FOR: Non-originality, deterioration through age, mistreatment.
CLASSIC CAR STATUS: * * Underrated as advanced product of a genius.

Chapman's first cars, built between 1948 and 1951, were rudimentary competition specials which showed great ingenuity despite their low cost. But his sixth design, the Lotus Mark 6 sports car, had one of the most advanced specifications in the world when its prototype first appeared in 1952. It was based around a spaceframe, which was outstandingly strong and light because Chapman knew all about stress factors through his full-time job as a civil engineer, during which he worked on metal bridges and similar stressed structures. The second significant factor of the Mark 6's design was that it used relatively soft suspension to achieve outstanding roadholding at a time when the science of making a car handle well centred mainly on stiffening the springs as much as possible.

A demand for replicas of the Mark 6 was met by selling the car in kit form so that customers could assemble it themselves with mechanical parts gleaned from other cars, notably Ford saloons. In this way, Chapman solved the problem of having no factory – only a shed behind his father's public house.

The cheap and simple Mark 6 was so popular, with more than 100 examples sold between 1953 and 1955, that Chapman was able to go into the business of making cars full-time. Initial development of the Mark 6 was in Mark 8 aerodynamic form for the racetrack in 1954, the Mark 7 having been a Formula 2 racing car project called the Clairmonte.

The beautifully streamlined Mark 8 body had been designed by De Havilland Aircraft Company aerodynamicist Frank Costin, whose brother, Mike, was Chapman's development engineer. The Mark 8, powered by a highly tuned M.G. engine, proved to be the fastest small sports racing car in Britain, before being developed into the even smaller and lighter Mark 9 in 1955. This car, which featured prominent aircraft-style fins on the tail to promote stability, was even smaller and lighter. It was to be powered by the new ultra-lightweight 1100 cc Coventry-Climax racing engine, with a larger Mark 10 version using Bristol's 2-litre competition unit.

Chapman's driving ability proved to be a great asset at the time as the newly-formed Team Lotus set its sights on the Le Mans 24-hour race. His Mark 9 led its class before being disqualified after an accident. Chapman's economical ideals also showed in the way the description 'Mark' was dropped for his next design, the 1956 Lotus Eleven (the Eleven was spelt out to avoid confusion with the Roman numeral II). This car was an improved version of the Mark 9, and was a fine illustration of Chapman's philosophy in that it was designed to accept a variety of engines; similarly, in the pursuit of efficiency, he made each component in his cars, or person in his company, responsible for

▽ *The Lotus Mark 6, produced by Colin Chapman, offered advanced technology at a very low price.*

as many functions as possible.

It was at this point that Chapman became involved in Grand Prix racing, not only as a driver, but also by laying out the chassis and suspension for the British Vanwall while Frank Costin designed the body. It was a further example, after the Clairmonte, of how Chapman capitalized on Lotus's design and engineering expertise to help finance future ventures. As Lotus worked more and more on its own single-seater and sports racing cars, Chapman developed the Mark 6 into the low-priced Lotus Seven. This model was produced from 1957, during the slack times that sometimes occurred between building racing cars.

SUMMARY

Plus points

- Excellent handling.
- Durability.
- Ease of maintenance and repair.
- Spartan appeal.
- Numerous events in which it can compete.
- Technically advanced spaceframe.

Minus points

- Very spartan.
- Cramped.
- Not very fast in a straight line.

ELITE

YEARS MADE: 1959–63.
ENGINE: In-line four-cylinder, 1216 cc, single overhead camshaft, typically 72 bhp at 6200 rpm.
FRAME: Integral.
PERFORMANCE: Typically 110 mph (177 km/h).
CURRENT PRICE GUIDELINE: *** Rapid recent escalation as model has at last appreciated.
PROBLEMS TO LOOK FOR: Cracking and general deterioration of glass-fibre bodyshell, poor paintwork, worn-out and broken mechanics, electrical problems, clumsy previous attempts at restoration, non-originality of wheels in particular.
CLASSIC CAR STATUS: **** Will eventually go higher.

One of Chapman's friends, accountant Peter Kirwan-Taylor, who had designed and built a special body for his own Lotus Mark 6, asked Colin if he could have one of the later chassis for the same purpose. If the design was successful, they could perhaps build a few replicas – providing there were enough people with the money to buy them. Chapman took the idea one stage further – why shouldn't they build such a car by the thousand to make even more money?

As Kirwan-Taylor sketched what was to be the most beautiful GT car in the world, Chapman showed another aspect of his genius. He dispensed with the chassis, incorporating the engine, transmission, steering and suspension mounting points in the bodyshell. This body would be made not from inherently heavy metal, but from the lighter glass fibre. Chapman saw no reason why this normally brittle material should not be made strong enough, with local reinforcements and the right stressing, to form what would be the world's first plastic car.

The theory was great, and Kirwan-Taylor's lines – honed to perfection by Frank Costin – were marvellous, but development of the Lotus Elite prototype, first seen in 1957, was painfully slow because glass-fibre technology was then in its infancy. The cash deposits generated by the first show model were enough to pay for the raw materials for the cars, but not enough, for instance, to finance either the new factory which they would need if they were to produce a large number of cars, or exports on which payment was delayed until after delivery, which could take several weeks. A third factor was that the workforce was still largely unskilled, even as production reached full swing in 1960. As a result, Elites

▽ ***The original Lotus Elite – the world's most beautiful GT car. Nigel Dawes Collection.***

△ ***The Lotus Elan sports car was developed through four distinct phases to the ultimate Sprint version.***

worked wonders on the track when they were attended by teams of top-class mechanics, but not as production cars. Problems caused by poor quality control reached massive proportions and nearly bankrupted Lotus.

It was then that Chapman showed financial genius. He turned the Elite into a kit car like the Mark 6 and, because no purchase tax was payable on cars that customers assembled themselves, slashed the price of what had been an expensive little machine. Eventually, by 1964, all stocks of unsold Elites had been cleared.

SUMMARY

Plus points

- World's most beautiful GT car.
- Superb handling and performance when in first-class condition.
- Excellent ride.
- Sensuous handling.
- Glass-fibre bodywork often resists corrosion.

Minus points

- Needs racing car-style maintenance.
- Delicate mechanical parts.
- Electrical system dangerous in original form (without fuses).
- Cramped interior.
- High costs of restoration, especially of bodywork.

ELAN SPRINT

YEARS MADE: 1971–3.
ENGINE: In-line four-cylinder, 1558 cc, twin overhead camshafts, 126 bhp at 6500 rpm.
FRAME: Fabricated steel backbone.
PERFORMANCE: 120 mph (193 km/h).
CURRENT PRICE GUIDELINE: * * Marginally undervalued in terms of performance.
PROBLEMS TO LOOK FOR: Corrosion in chassis, cracking and general deterioration of bodywork, poor paintwork, worn-out mechanical parts (especially engine's water pump), mistreatment, non-originality with inferior replica parts.
CLASSIC CAR STATUS: * * Will rise gradually.

As his racing cars became increasingly successful, Chapman was determined to capitalize on the Lotus name's competition prestige with his road cars. Profits on the Seven were limited by its low price and production costs which could not be cut because it was based around the labour-intensive spaceframe. One of the main problems which had forced up the cost of the Elite was the price of its competition-based Coventry-Climax engine. The cheap new Ford Anglia small car engine had been developed to give high power outputs in competition by a new firm, Cosworth. Chapman therefore decided to use the Cosworth engine as a base for a new sports car unit. He then hired freelance Harry Munday to design a Jaguar XK-style twin-camshaft cylinder head conversion for the larger Cortina development of the Ford engine, which gave it more power at a lower cost.

At the same time, new suspension developed on Lotus racing cars and attempts at making a cost-cutting one-piece glass-fibre body were being tested by development engineer Ron Hickman. He had no road car chassis to run as a test bed, so he made up a backbone frame, based on that of his mother's Mercedes 300 saloon, to try out the new components. Chapman then realized that this test set-up was basically a very good idea, and his second quantity-production sports car, the Elan – still front-engined like the Elite – appeared in 1962.

The Elan was a far greater success than its predecessor, partly because its backbone chassis meant that it did not need a stressed roof for strength and could be produced in the more popular open form, and partly because Lotus was becoming more experienced in producing cars in quantity! The two-seater Elan was developed through four distinct series, with an optional hard-top version from 1966, and even more powerful variants of the Lotus twin-cam engine, for the following ten years. During 1967, a wider, more upmarket version, with two small rear seats, called the Elan +2, was added to the range. In February 1971, the basic model was renamed the Elan Sprint, but it was now powered by a big-valve version of the original twin-cam engine and had distinctive two-tone livery. Eventually, all models in the Elan series were sold in ready-to-drive form, as the kits offered only small profits.

SUMMARY

Plus points

- Superb handling and performance, especially acceleration, when in good condition.
- General resistance to corrosion of glass-fibre bodywork.
- Good availability of spares.
- General ease of restoration.

Minus points

- Delicate chassis and suspension.
- Typical corrosion in (replaceable) chassis.
- Needs racing car-style maintenance.
- High costs of bodywork restoration.

EUROPA SPECIAL

YEARS MADE: 1972–5.
ENGINE: In-line four-cylinder, 1558 cc, twin overhead camshafts, 126 bhp at 6500 rpm.
FRAME: Steel backbone.
PERFORMANCE: 123 mph (198 km/h).
CURRENT PRICE GUIDELINE: * *
Generally underrated when in good condition.
PROBLEMS TO LOOK FOR:
Mistreatment, corrosion, accident damage, mechanical deterioration.
CLASSIC CAR STATUS: * * Set to appreciate.

After the Lotus-Cortina, produced by Lotus using Ford bodyshells from 1963-6, and the Cortina-Lotus, produced by Ford until 1968, Chapman still dreamed of producing a small sports machine in quantities that would make him a major road car manufacturer. This had been his initial intention with the Elan, but it had become more expensive, partly because its bodyshell could not be made in just one piece. Chapman was also tantalized by the obvious trend towards mid-engined sports cars, demonstrated by the Ford GT40 project that had gone to his racetrack rivals, Lola (see page 75).

So once more Chapman tried to create the one-piece bodyshell for a cheap, new, mid-engined car, to be powered by a Renault engine. The French firm had initially been approached because it had the only suitable transmission for the new car, but it ended up selling the engine, too, and at a much-reduced price since the deal would give Renault some good publicity, as did its involvement with Alpine (see page 19). It looked as though Britain would soon enter the Common Market, so the new car, introduced in 1967, was called the Europa.

It featured a backbone chassis like the Elan and racing car-style suspension. This, combined with the fact that its body could not, after all, be made in one piece, meant that the Europa's price rose to near that of the Elan. In fact, due to the high insurance premiums for this car, it soon became more expensive as it had to be given a detachable chasis, instead of having it moulded in, to reduce repair costs. Once the Europa's price had risen, it became apparent that its performance and specification would have to be improved. The Alpine's sales would be badly hit if the Lotus car received really powerful Renault engines, so the Lotus twin-cam was substituted in 1971 which – with detail refinements – made the Europa a much more desirable car.

After his difficulties in obtaining a more powerful Renault engine, Chapman was anxious to avoid becoming too dependent on other manufacturers, and decided to build his own power unit. This needed a lot of capital investment, but was achieved partly with the help of the Jensen-Healey project (see page 92), which used his new twin-overhead-camshaft in-line four-cylinder engine from 1972.

SUMMARY

Plus points
- Performance.
- Wonderful sensuous handling.
- Excellent availability of spare parts.

Minus points
- Delicate construction.
- High level of maintenance vital.

▽ The Lotus Europa, one of the first mid-engined production cars in 1966, developed into the Special in 1972.

ESPRIT TURBO

YEARS MADE: 1980 to date.
ENGINE: In-line four-cylinder, 2174 cc, twin overhead camshafts, 210 bhp at 6000 rpm.
FRAME: Fabricated steel backbone.
PERFORMANCE: 152 mph (245 km/h).
CURRENT PRICE GUIDELINE: ** to *** According to age.
PROBLEMS TO LOOK FOR: Lack of maintenance, mistreatment.
CLASSIC CAR STATUS: *** Will appreciate.

By 1975, Colin Chapman had despaired of ever making much money from cheap sports cars, and he replaced the Elan and the Europa with a new Elite, using his own engine, to a format based on the very successful Reliant Scimitar GTE sporting estate. This was developed at the same time into a striking Giugiaro-styled wedge-shaped mid-engined GT car called the Esprit, and a two-plus-two variant called the Eclat, all using a development of the Europa backbone chassis. By 1981, the Esprit had reached its ultimate form in the Esprit Turbo – one of the world's fastest GT cars. A V8 version of the Esprit was waiting in the wings when, in 1982, Chapman died suddenly. By then, much of Lotus's income was coming from design and development for other manufacturers, and this asset was realized when it was taken over by General Motors to continue the research that Chapman had pioneered.

SUMMARY

Plus points

- Tremendous performance.
- Superb turbo installation.
- Excellent handling, especially on difficult roads.
- Still-stunning Giugiaro styling.

Minus points

- Restricted rear vision, poor visibility.
- Bad ventilation.
- Needs detailed maintenance.
- High restoration costs on glass-fibre bodywork.
- Possible corrosion problems with chassis after approximately six years.
- Delicate suspension.

MARCOS

The Marcos marque has provided some of the most ingenious and forward-thinking machinery to be seen on road and track. The bodies of early examples were made of marine-quality plywood which proved not only very light, but rigid too, and helped endow the cars with excellent handling. Their appearance, however, was a disaster. Later, stylist Dennis Adams replaced aerodynamicist Frank Costin, and the stunning low-line Marcos GT was created in 1963 with such an individualistic outline that it has survived in production into the 1980s.

MARCOS GT

YEARS MADE: 1964–71, 1981 to date.
ENGINE: Typically Volvo in-line four-cylinder, 1778 cc, overhead valves, pushrods, 115 bhp at 6000 rpm, or Rover V8-cylinder, 3523 cc, overhead valves, pushrods, 180 bhp at 5200 rpm.
FRAME: Wooden or tubular steel.
PERFORMANCE: Typically 115 mph (185 km/h) to 135 mph (217 km/h).
CURRENT PRICE GUIDELINE: ** to *** according to age and specification.
PROBLEMS TO LOOK FOR: Mistreatment.
CLASSIC CAR STATUS: *** Truly striking appearance.

One of the most popular occupations among motoring enthusiasts in Britain immediately after the Second World War was building special cars for competition. The parts for these specials came mainly from old Austin Sevens as they were very cheap and available in reasonably large quantities (see page 26). Competitions were organized specially for these cars by the 750 Motor Club, which took its name from the Austin Seven's engine capacity of 750 cc. The club also coordinated various activities. The regulations for its competitions were framed to place a premium on ingenuity within a very tight budget, and this encouraged brilliant young designers who were working on a shoestring, such as Colin Chapman, whose first Lotus cars were constructed for these competitions, and Eric Broadley, his some-time rival with Lola racing cars.

The 750 Motor Club's activities also led to small firms being set up to supply specialized parts for these cars. One such firm was Dante Engineering, which employed a young engineer, Jem Marsh, in the mid-1950s. Marsh soon realized that he could do better and left in 1957 to start his own business, Speedex, supplying accessories for Austin Seven specials – rather like Karl Abarth's operation with Fiats in Italy (see page 8).

▽ The Lotus Esprit Turbo, developed along lines pioneered by the Europa, has become a great supercar.

△ ***Still in action in 1986, the classic Marcos GT which created a sensation in the 1960s.***

Two years later, at a 750 Motor Club meeting, Marsh met aerodynamicist Frank Costin, who expounded his theories on the advantages of using wood in car construction. Between them, the pair mapped out an ultra-lightweight wooden version of the popular Lotus Mark 6 (see pages 108-9), with the added advantage of a fixed top to give all-weather protection and improved aerodynamics. Costin was to be responsible for the plywood monocoque (built like the fuselage of the Mosquito bomber), while Marsh would supply the running gear.

It was during the summer of 1959 that the first Marcos (from *Mar*sh and *Cos*tin) emerged. It had a pointed nose, cycle-type front wings, gullwing doors and a wide-hipped tail, displaying brilliant aerodynamics but a rather ungainly styling which earned it the nickname, 'Ugly Duckling'.

Frank Costin quit the project at this point. Soon afterwards the Marcos acquired a formidable performance with a new Ford Anglia overhead-camshaft engine developed by the Cosworth tuning firm set up by Frank's brother Mike, and former Lotus employee Keith Duckworth. Satisfied customers included future world champion racing driver Jackie Stewart.

Then a new spaceframe chassis was designed by competition specialist Dennis Adams, who also revised the bodywork to improve the model's appearance and make it more saleable. An open version, the Marcos Spyder, was also built in 1962, to save the cost of making the gullwing doors.

Marcos's customers wanted a hardtop version for GT racing, however, so Adams then redesigned the Spyder in 1963 as the GT Fastback (so named because of its long sloping tail).

After finishing the GT Fastback, Adams started work on an ultra-low GT car of very futuristic appearance. This reverted to a wooden chassis with glass-fibre body panels and it was powered by a Swedish Volvo engine. Its light weight and nimble handling again ensured an excellent performance, but its appearance was so stunning, with a Grand Prix-style reclining driving position, that sales of the new Marcos GT really took off. In the next few years the GT was given a wide variety of engines and running gear, and in 1969 a more economical tubular steel chassis was introduced. Although output ceased in 1971 due to financial problems, ten years later the demand for secondhand GTs was such that Marsh was able to put the car back into production.

A Marsh project of the mid-1960s used the cheap running gear from the BMC Mini saloon car in a Lotus Elite-style glass fibre bodyshell (see pages 130 and 109). The result appeared in 1965 as the Mini-Marcos. The Marcos range was highly successful, and the company prospered.

Meanwhile, a Marcos Mantis sports racing car project had led to production of a new four-seater GT car of that name which pre-dated Lotus's 1975 Elite by four years. However, over-expansion, the costs of setting up a new factory, and an export cashflow crisis forced the company to close.

In 1984, three years after demand for the Marcos GT had brought Marsh back into the car manufacturing business, Adams was asked to restyle the body slightly to accommodate a larger Rover V8 engine for a model called the Mantula. It was at this point that Speedex re-emerged after 30 years to supply specialist parts again, and – in 1986 – an open version of the Mantula was marketed to provide another Marcos Spyder.

SUMMARY

Plus points

Stunning appearance.

Excellent performance.

Good handling.

Excellent availability of spares.

Durable production-based mechanics.

Minus points

Restricted visibility from reclining driving position.

Hard ride by modern standards.

MASERATI

The renowned Italian Maserati company produced only racing cars from the time when it was founded in 1926 by four brothers of that name until the beginning of the Second World War. Like many producers of racing cars, they were far better at building them than at making money!

The Maserati brothers had some rocky times in the early years and had to look for some outside finance. This was forthcoming from industrialist Count Omer Orsi in 1937, and in return the Maserati brothers signed a ten-year contract giving him their name and services. All went well until just before the war, when they had to switch production from their beloved Bologna to Orsi's headquarters in Modena, in the interests of efficiency. From that point, the relationship deteriorated as the brothers realized that they had lost their independence. On the day that the contract expired, the Maserati brothers returned to Bologna to set up a new business producing racing cars under the name OSCA, leaving Orsi with their name and designs.

A6GCS

YEARS MADE: 1947–55.
ENGINE: In-line six-cylinder, 1985 cc, twin overhead camshafts, typically 170 bhp at 7300 rpm.
FRAME: Tubular.
PERFORMANCE: Typically 146 mph (235 km/h).
CURRENT PRICE GUIDELINE: **** Rapid recent appreciation with advent of Mille Miglia-style rallies.
PROBLEMS TO LOOK FOR: Non-originality, mechanical wear, mistreatment, parts missing.
CLASSIC CAR STATUS: *** All-enveloping versions are among the most beautiful sports racers.

In 1946, the astute Orsi developed the successful pre-war 6CM racing car into a dual-purpose machine for the new Formula 2 or, with vestigial bodywork, for sports car racing – as Ferrari had done with the 166. The design of this 1.5-litre in-line six-cylinder car – called the A6/1500 – with its tubular chassis and coil spring suspension, was typical of contemporary Italian competition machines. It was very expensive and sold only in small numbers until, with the A6G/2000, the power of its single-overhead-camshaft engine was increased by taking out the capacity to 2 litres. This model was to be produced from 1951 alongside Maserati's existing racing cars. Bodybuilders who used the A6G/2000 base included Pininfarina, Frua and Vignale.

A similar chassis and running gear was used with a more powerful twin-overhead-camshaft cylinder head (by former Ferrari designer Giaocchino Colombo) in the outstandingly attractive sports racing A6GCS.

△ *Maserati's first attempt at producing a sports car developed into the lovely A6GCS.*

SUMMARY

Plus points
- Sheer beauty of all-enveloping body.
- Very high performance.
- Safe handling for amateurs.
- Numerous events in which it can compete.

Minus points
- Far from durable mechanics.
- High costs of restoration.
- Needs a racing car's maintenance.

GHIBLI

YEARS MADE: 1966–73.
ENGINE: V8, twin overhead camshafts, 4709 cc or later 4931 cc, 330 bhp at 5500 rpm or 335 bhp at 5500 rpm.
FRAME: Tubular.
PERFORMANCE: 165mph (265 km/h), or 174 mph (278 km/h).
CURRENT PRICE GUIDELINE: **** All-time classic appearance, especially of Spyder version.
PROBLEMS TO LOOK FOR: Corrosion in body.
CLASSIC CAR STATUS: ***** One of the best-looking road cars ever built.

Maserati's reputation was at a record height in 1956 because of the success of its 250F Grand Prix car. Any new Maserati production car could be sold at a higher price because the marque's reputation had been established with glamorous single-seaters.

Chief engineer Giulio Alfieri, therefore, reworked the Jaguar-like twin-overhead-camshaft straight-six-cylinder engine of the 250F's close relative – the 350S sports racing car – for larger-scale production in a grand touring car. Underneath the bonnet, it was given two sparking plugs per cylinder (as many as a V12 Ferrari) and three Ferrari-like Weber carburettors. The running gear was bought in from a variety of component suppliers abroad, but the bodywork of the 3500GT was pure Italian. By 1958, the first serious Maserati road car – in open or closed form – was selling in greater quantities than the A6G had in the previous seven years. This was the first in a line of impressive models which culminated in the 1967 Ghibli supercar.

When Maserati stopped running its own racing cars in 1957, it was left with a small, but valuable, stock of very powerful 5-litre V8 engines. As a result, the 3500GT chassis was revised to accept these power units in a model called the 5000GT. This was produced in only very small quantities at a very high price with a choice of exclusive bodywork – like Ferrari's Superfast (see page 65).

The advent of Jaguar's 150 mph (240 km/h) E-type in 1961 (see page 89) forced Maserati to adopt fuel injection for the 3500GT like that used on the 5000GT, so that the engine's power output would be increased to give it a performance comparable to the British car's. Two years later the 3500GT also received disc brakes on all four wheels and was restyled by Vignale as the Sebring, recalling victories by the Bird-cage at the American circuit.

More power was extracted from the 3500GT's engine by increasing its capacity to 3.7 litres for a new top model, the 1963 Mistral. The car also had more attractive bodywork by Frua. This short-wheelbase model, distinguished by a 'goldfish bowl' rear window and a nose with a prominent chrome-plated bumper, became one of Maserati's most successful cars.

In 1963, Maserati caused a sensation by following the Mistral with a four-door supercar, the Quattroporte, by Frua. It featured a much-revised chassis (with de Dion rear suspension until 1967) and a detuned 4.1-litre version of the V8 engine which had been used until 1964 in the 5000GT. This car proved to be a great inspiration for Lamborghini's Espada (see page 98), but even when it was fitted with a more powerful 4.7-litre engine in 1969, it proved too conservatively styled to be as successful as Maserati had hoped.

In the meantime, Maserati had decided 'to back both camps' in 1966 by introducing a new American-oriented two-plus-two car by Vignale, the two-door Mexico (harping back to the Carrera Panamericana road race), using the V8 engine in 4.7-litre form to give it a performance comparable with that of the Mistral. By this time, however, stronger competition from Ferrari road cars (and, to a certain extent, from Lamborghini and Iso) was limiting Maserati's expansion, so the Mexico was not a big-seller.

Mexico sales also suffered from stiff in-house competition as, in 1967, Maserati introduced the sensational Ghibli two-seater, named after a Saharan wind and designed by Giorgetto Giugiaro for Ghia. This was immediately proclaimed as one of the most stunning supercars ever built, especially in its open spyder form. Sales were further boosted when the Mistral was eventually phased out in 1970, and the Ghibli's performance increased to an even higher pitch as the SS (for Super Sport), with a 4.9-litre V8 engine.

SUMMARY

Plus points

- Fantastic appearance.
- Comfort.
- Superb reliability of mechanical components.
- Extremely high performance.
- Very durable if protected from damp climate and properly maintained.

Minus points

- Tendency to corrosion in body.
- High restoration costs.
- Troublesome electrics.

▽ ***The Maserati grand touring cars of the 1960s were highly attractive, especially Giugiaro's Ghibli.***

BORA

YEARS MADE: 1971–9.
ENGINE: V8, 4709 cc, twin overhead camshafts, 330 bhp at 6000 rpm.
FRAME: Integral.
PERFORMANCE: 175 mph (280 km/h).
CURRENT PRICE GUIDELINE: * * * Undervalued against contemporary Ferraris.
PROBLEMS TO LOOK FOR: Corrosion, mistreatment.
CLASSIC CAR STATUS: * * * * Very rapid and solid performer.

Maserati became closely associated with Citroën as part of a deal to produce the V6 version of the basic V8 for the French company's new SM saloon, leading to a full alliance in 1970.

Citroën's influence and investment could immediately be seen by the way that the Quattroporte was phased out to be replaced by the Mexico, which offered the more economical 4.1-litre engine as an option, in company with a restyled two-plus-two, the Indy (commemorating an early American success). This car was the first Maserati road car to use unitary construction.

Giorgetto Giugiaro, who had left Ghia to start his own consultancy, Ital Design, was then called in to produce a completely new body for a mid-engined Maserati supercar, even though the Ghibli was still selling well. This project reached fruition in 1971 as the Bora (an Adriatic wind), and the coupé showed a distinct family resemblance to the Indy. It was powered by the 4.9-litre V8 engine and was the first Maserati to feature coil spring and wishbone suspension. Inside the car was evidence of Citroën's influence, particularly in the use of the hydraulic brake system from the SM.

Soon after the Bora appeared, it was joined by the visually similar Merak (yet another wind), using a 3-litre version of the V6 engine, simultaneously introduced into the Citroën SM. The Merak's appeal was further heightened by its two tiny rear seats (in space liberated by the removal of two of the V8's cylinders) which put it into contention with Porsche's 911S, Ferrari's Dino and Lamborghini's Urraco (see pages 144, 69 and 99). The Bora continued in production until 1979 and the Merak until 1981.

SUMMARY

Plus points

Immense performance.

Mechanical durability.

Minus points

Not so beautiful as earlier front-engined GTs.

Corrosion in body.

High restoration costs.

▽ ***The Bora broke with Maserati tradition – it had a mid-engine and modern coil spring suspension.***

KHAMSIN

YEARS MADE: 1974–8.
ENGINE: V8, twin overhead camshafts, 4931 cc, 335 bhp at 5500 rpm.
FRAME: Unitary.
PERFORMANCE: 140 mph (225 km/h).
CURRENT PRICE GUIDELINE: * * *
Vastly underrated supercar by Ferrari standards.
PROBLEMS TO LOOK FOR: Corrosion in body.
CLASSIC CAR STATUS: * * * * *
Extraordinary styling really does grow on you.

Legendary bodybuilders Bertone, who had until then been fully occupied with work for Lamborghini and Iso, were called in by Maserati to design a large two-plus-two replacement for the Indy. Like the Bora and Merak, this new model represented a new generation of Maserati road cars using independent rear suspension, earlier models having survived with less sophisticated, but very reliable, solid rear axles. The new car, introduced in 1973 with the 4.9-litre engine, was called the Khamsin, after a hot desert wind. Again, a strong Citroën influence could be seen and the Khamsin also bore a family resemblance to Bertone's Lamborghini Urraco (see page 99).

In the uncertainty of the world economic climate leading up to the 1973/4 energy crisis, the Ghibli SS, Indy and Mexico, of the traditional cars, continued to be marketed, while production of the new Bora, Merak and Khamsin built up before the crisis, combined with American safety legislation, badly hit sales of supercars.

△ ***Maserati's Khamsin represented a new generation of road cars and featured advanced Citroën technology.***

SUMMARY

Plus points
- Unusual appearance.
- Comfort.
- Modern standards of ride and handling.
- Superb reliability of mechanical components.
- Sophistication of Citroën hydraulics.
- High performance.

Minus points
- Tendency to corrosion in body.
- High restoration costs.
- Troublesome electrics.

MERCEDES

The German Mercedes cars can trace their origins to Karl Benz, who in 1885 produced the world's first motorized carriage for use on the road. Since then they have maintained a superb reputation for quality, and frequently for the highest technology, too.

Benz's greatest rival in the early days was another German, Gottlieb Daimler, whose cars, which would be given the name Mercedes after the daughter of sales director Emil Jellinek, soon outstripped the Benz products technologically.

The naturally conservative Karl Benz was stung into action, and until the First World War the rival marques Mercedes (still produced by the parent firm Daimler) and Benz struggled for supremacy. Mercedes' crowning glory was a fantastic overhead-camshaft Grand Prix machine of 1914 which led to a revolution in engine design.

During the war both Daimler and Benz became industrial giants, but in the early 1920s they had to struggle – against runaway inflation this time – until they were forced to pool their resources and become Daimler-Benz in 1926, producing Mercedes-Benz cars, trucks and aero-engines.

Early chief engineers included Paul Daimler, son of Gottlieb, and a professor who was to become legendary, Ferdinand Porsche. As the Benz designs were developed to provide the workaday Mercedes models, the original Mercedes line was extended to include much more glamorous machinery.

SSK

YEARS MADE: 1928–32.
ENGINE: In-line six-cylinder, 7069 cc, single overhead camshaft, supercharged, 200–250 bhp at 3300 rpm.
FRAME: Twin girders, cross-braced.
PERFORMANCE: 115 mph (185 km/h) to 125 mph (201 km/h).
CURRENT PRICE GUIDELINE: *****+ One of the most valuable cars in the world.
PROBLEMS TO LOOK FOR: Non-originality, engine damage.
CLASSIC CAR STATUS: ***** Gilt-edged investment.

The first supercharged Mercedes sports cars – the 6/25/40 and the 10/40/65 – had been Paul Daimler's last great work before he retired in 1922. Later that decade they were to evolve into the SSK, one of the most valuable cars in the world today. The first figure in the designations represented the German taxable horsepower, the second the normal power output of the engine, and the third showed the total output when the mechanical supercharger was engaged with a spine-chilling scream!

In ordinary circumstances, the engine operated under atmospheric pressure, but when maximum acceleration or full speed was needed, the supercharger was engaged by pushing the throttle to the floor for a few seconds. Woe betide anybody who kept their foot down hard for too long . . . they risked blowing the top off their engine as the supercharger boosted the internal pressure too high!

Porsche's first work when he succeeded Paul Daimler was to develop these cars using technology learned from a 2-litre supercharged eight-cylinder racer of 1923.

The specifications of the resulting 15/70/100 and the massive 24/100/140 were typical of the biggest and most expensive sporting cars built in the 1920s. They had long-wheelbase channel section chassis with solid axles front and rear and very stiff springs. Their roadholding depended almost entirely on their great weight and flexing chassis, with the ride made tolerable only by the length of the wheelbase and the depth of their upholstery.

That said, the engine was a magnificent piece of work that became more powerful by the day. It needed to be: some of the splendid coachwork used on these cars made them the heaviest on the road.

Profit margins on these top-of-the-range cars were high, but production was restricted because so much work went into their beautiful construction.

One of the main problems with the 24/100/140 was that its handling could be alarming when extended. As a result, reducing its weight became a greater priority than extracting more power from its already formidable engine.

One of the first changes came in 1926 when the chassis was shortened for a new model which bore the same designation, but was called K for *kurz* (short). This reduced the weight of the rolling chassis alone to around 1½ tons (1525 kg) in sporting form, 2 tons (2035 kg) for a touring body, or 2¼

▽ *The awesome Mercedes SSK has emerged as the marque's most formidable pre-war road car.*

△ ***The Grosser Mercedes, made for heads of state and industry, leaned on sophisticated truck practice.***

tons (2300 kg) in long-wheelbase form.

Subsequent evolution included a 26/120/180S with the shorter wheelbase, a far lower chassis for better handling, and a larger 6.8-litre engine. This much lighter S-for-Sport model weighed only 1¾ tons (1900 kg) complete with a fairly spartan body.

The ultimate development on this theme was the SSK of 1928, so called because it was the Super Sports Short. The SSK had an even shorter wheelbase and bigger 7.1-litre engine with a higher compression ratio, a combination which gave it the technical designation 27/170/225 and the ability to outrun even the previously all-conquering big Bentleys. Eventually this car was developed through the 27/180/250SSK to the even lighter and more powerful 27/240/300SSKL of 1930.

Unfortunately, such power outputs were achieved at the expense of reliability, and the original SSK, with its reasonably durable engine, emerged as the classic.

SUMMARY

Plus points

Rarity value.

Savage performance.

Awe-inspiring appearance.

Wonderful sound of supercharger in action.

Beautifully built.

Minus points

Engine can be fragile if misused to extremes.

Wild handling.

High restoration costs.

GROSSER MERCEDES

YEARS MADE: 1930–8.
ENGINE: In-line eight-cylinder, 7655 cc, overhead valves, pushrods, supercharged, 200 bhp at 2800 rpm.
FRAME: Twin girders, cross-braced, later tubular.
PERFORMANCE: 86 mph (138 km/h).
CURRENT PRICE GUIDELINE: **** All-time giant.
PROBLEMS TO LOOK FOR: Non-originality, false claims to history.
CLASSIC CAR STATUS: ***** Never-to-be-forgotten giant.

In the late 1920s, Mercedes was equipped to handle almost any demand for a high-quality car, except perhaps from captains of industry or heads of state who wanted an even more imposing carriage.

Former Benz engineering chief Hans Nibel took over from Porsche, and his

△ ***The magnificent Mercedes 540K was one of the heaviest, most powerful and luxurious pre-war supercars.***

first model in 1930 was a truly monumental machine, called the Grosser Mercedes because it was so big. It was largely based on the company's best trucks, and so was technically not as advanced as the sports cars. However, it boasted a massive 7.7-litre supercharged engine which was enough to haul along its 3½ tons (3500 kg) at a remarkable speed, given the load.

Naturally this was a favourite car of Adolf Hitler and high-ranking members of the Nazi party, and was even used by Emperor Hirohito of Japan until the mid-1960s.

SUMMARY

Plus points

Sheer size.

Enormously durable.

Magnificent appearance.

Feeling of invincibility.

Minus points

Very heavy steering.

Almost every one found is claimed to be a *Führerwagen*.

540K

YEARS MADE: 1936–42.
ENGINE: In-line eight-cylinder, 5401 cc, overhead valves, pushrods, supercharged, 180–200 bhp at 3400 rpm.
FRAME: Twin girders, cross-braced.
PERFORMANCE: 106 mph (170 km/h) to 112 mph (180 km/h).
CURRENT PRICE GUIDELINE: *****+ Top-line investment.
PROBLEMS TO LOOK FOR: Non-originality.
CLASSIC CAR STATUS: ***** Marvellous asset.

As the nimble Bugattis and Alfa Romeos began to take over from the SSK in sports car events, and Mercedes concentrated on Grand Prix racing, the top models became increasingly exotic touring cars. The chassis were unusually advanced for their day in that they used independent suspension all round to provide a far more civilized ride, even if the handling could be difficult due to the relatively unsophisticated swing axles at the back.

The new suspension was then transferred to the normal cars. The first of these far higher-powered and more luxurious supercars became the 380K (for *Kompressor* – supercharger), before the engine's capacity was increased to 5 litres for the 500K in 1934.

Far superior de Dion rear suspension was then used on prototypes of the even faster 540K in 1936 before the Mercedes management decreed that all production 540Ks should stick to the cheaper swing axle rear suspension so that their normal models did not appear to be inferior. Many versions of the 540K were made until the enforced cessation of production in 1942. A lightweight version clocked 112 mph (180 km/h) in 1937, although most 540Ks carried similar heavyweight bodies to those used on the 500K. Later versions had V12 engines, and also featured the five-speed gearbox used on the last of the Grosser Mercedes. Meanwhile the more basic cars were developed with a superior tubular chassis to make their suspension safer.

SUMMARY

Plus points
Rarity value.

Magnificent engineering.

Extraordinary performance for a touring car of its day.

Short-wheelbase lightweight roadster even more extraordinary.

Durability.

Minus points
Steering very heavy especially at low speeds.

Lack of agility.

300SL COUPÉ

YEARS MADE: 1954–7.
ENGINE: In-line six-cylinder, 2996 cc, single overhead camshaft, 215 bhp at 5800 rpm.
FRAME: Tubular spaceframe.
PERFORMANCE: 165 mph (265 km/h).
CURRENT PRICE GUIDELINE: * * * * Set to appreciate.
PROBLEMS TO LOOK FOR: Accident damage, mistreatment, undershield missing.
CLASSIC CAR STATUS: * * * * *
Wonderful investment.

The earlier sports racing cars were developed into the sensational gullwing-doored 300SL road car of 1955. Its specification included such advanced developments as fuel injection, instead of conventional carburettors, which made it even more powerful than the earlier racing versions! The only problem, however, was that the 300SL still had swing axle rear suspension, which worked well in the hands of expert drivers, but could be dangerous when used by the inexperienced.

The 300SL was modified in 1957 to improve the handling and was given open bodywork to avoid occupants being trapped by the gullwing doors of the fixed-head version.

SUMMARY

Plus points
Tremendous performance.

Beautifully built.

Wonderful appearance.

Minus points
Swing axle geometry and missing aerodynamic underbody can promote lethal handling.

▽ ***The gullwing-doored Mercedes 300SL fixed-head coupé. Owner: Peter Parsons.***

300SLR

YEAR MADE: 1955.
ENGINE: In-line eight-cylinder, 2982 cc, desmodromic valvegear, 296 bhp at 8000 rpm.
FRAME: Tubular spaceframe.
PERFORMANCE: 180 mph (290 km/h).
CURRENT PRICE GUIDELINE: * * * * * + Best investment of all if you could persuade Daimler-Benz to part with one!
PROBLEMS TO LOOK FOR: Competition wear and tear, lack of maintenance.
CLASSIC CAR STATUS: * * * * * All-time great road racer.

Daimler-Benz's factories were badly damaged during the Second World War. It was not until 1948 that car production got into full swing again, and it was 1952 before the company could think about producing anything glamorous. However, when lightweight racing versions of the top-line 300S saloon were constructed, they proved so reliable that they won four out of the five major events they contested!

This encouraged Mercedes to build a new Grand Prix car to show that its technology was still the best in the world. The resultant W196 spawned a 300SLR (for *Sport Leicht Rennsport*) sports racing version and these became two of the most formidable competition cars in the world, winning the Mille Miglia in 1955 and the world single-seater title in the same year. The 300 SLR closely resembled the W196, of course, although the spaceframe chassis was made a little wider to make room for a second occupant. One other obvious difference lay in the SLR's ingenious air brake that popped up behind the driver's head when he needed to boost the braking power.

SUMMARY

Plus points
Rarity value.

Exotic specification.

Competition history.

Beautiful appearance.

Performance.

Minus points
All confined to museums with only occasional demonstrations outside.

△ ***The 1955 Mercedes 300SLR Mille Miglia winner. Its number 722 denotes its 7.22 am starting time.***

230SL

YEARS MADE: 1963–7.
ENGINE: In-line six-cylinder, 2281 cc, single overhead camshaft, 150 bhp at 5500 rpm.
FRAME: Integral.
PERFORMANCE: 124 mph (200 km/h).
CURRENT PRICE GUIDELINE: * * Undervalued on styling.
PROBLEMS TO LOOK FOR: Corrosion, wear and tear.
CLASSIC CAR STATUS: * * * Set to appreciate.

In 1963, the basic 220SE (for *Super Einspritz* – Super with fuel injection) saloon was developed into the 230SL, a slightly more powerful grand touring car. It had all the comforts of one of the top saloons, such as power-assisted steering and the option of automatic steering, but it still had the appearance of a sports car because of its exceptionally lithe styling. It also had disc brakes for the first time on any Mercedes car. As the engines of Mercedes' top saloon were increased in capacity, so the GT model followed suit, initially as the 250SL and then as the 280SL.

The 230SL version stayed in production until 1967.

SUMMARY

Plus points
Beautiful pagoda top styling.
Comfort.
Well made.
Durable if treated properly.

Minus points
Tendency towards extensive corrosion.
Poor gearbox ratios.
High maintenance and restoration costs.
Expensive parts.

MERCEDES-BENZ 600

YEARS MADE: 1964–81.
ENGINE: V8, 6330 cc, single overhead camshaft, typically 250 bhp at 4000 rpm.
FRAME: Integral.
PERFORMANCE: 120 mph (193 km/h).
CURRENT PRICE GUIDELINE: * * to * * * * Depending on age and condition. Set to appreciate.
PROBLEMS TO LOOK FOR: Lack of detailed preventive maintenance, neglect.
CLASSIC CAR STATUS: * * * Set to appreciate.

As each basic saloon bettered its predecessor, there was a revival of demand for a car like the Grosser Mercedes. This was met, in 1964, by the 600, one of the most luxurious cars ever made, in two forms: the normal saloon could carry five or six passengers in great comfort, while a long-wheelbase Pullman version was also available with a six-door body that could carry seven or eight passengers in railway carriage-like opulence. Both versions of the 600 were powered by a superb new V8 cylinder engine which gave them a quite exceptional performance for large and luxurious limousines. They became two of Mercedes' longest-serving models, remaining in production until 1981.

△ ***The pagoda-style hardtop of the Mercedes 230SL became a prized feature. Owner: David Prior.***

SUMMARY

Plus points

Sheer size.

Extreme luxury, grandeur.

Extraordinary handling and ride for size.

Durability, given a high standard of preventive maintenance.

Minus points

Full of complex electrical and hydraulic gadgets to go wrong.

Corrosion in damp climates.

Very high maintenance costs.

Vast cost of restoration.

Massive fuel consumption.

▷ ***The Mercedes 600 six-door Pullman limousine had opulent coachwork. Owner: John Turner, Bruton's of London.***

300SEL 6.3

YEARS MADE: 1967–72.
ENGINE: V8, 6330 cc, single overhead camshaft, 250 bhp at 4000 rpm.
FRAME: Integral.
PERFORMANCE: 138 mph (221 km/h).
CURRENT PRICE GUIDELINE: * Set to appreciate dramatically.
CLASSIC CAR STATUS: ** Very high performance and level of comfort and luxury for its day.

In the late 1960s there was still a demand for a very fast big saloon, so the 600's 6.3-litre V8 engine was fitted to the normally 3-litre top-of-the-range 300SE model to endow the resultant 300SEL 6.3 of 1968 with a truly astounding performance. It was, in fact, so impressive that a works team was entered in the Spa 24-hour saloon race, although problems with overstressed tyres led to the cars being withdrawn after the practice session.

A new 3.5-litre V8 engine was developed to appear in the saloons and the open 350SL and fixed-head 350SLC from 1971. Both had striking wedge-shaped bodies and new running gear, including semi-trailing arm rear suspension: at last Mercedes had brought its handling up to date!

One of the reasons for sticking to swing axles for so long had been the company's desire to produce something for everybody in the quality motoring world.

The amount of engineering time spent by Mercedes in developing the range was extraordinary. Although diesel cars made up the bulk of production, there was an incredible number of permutations which could be made on petrol-engined themes.

There were basic carburettor four-cylinder engines, normally aspirated and fuel-injected sixes from 2.3 and 2.8 litres, and injected 3.5 and 6.3 litres, which could be switched between saloon, coupé or convertible bodyshells with automatic or manual transmissions and seemingly endless variations on trim.

These variations were augmented still further in 1972 with a new top line of super saloons to replace the 300SEL 6.3, called the S class. Such machines were as near perfection as Mercedes could make them.

Although the 350 models coped quite well with America's ever tougher and more power-sapping exhaust emission regulations, Mercedes went one step further in 1973 by introducing a 450 series of 4.5-litre V8s to ensure that no performance was lost.

But still Mercedes was not satisfied and continued its quest for perfection in the 1970s with research into how the S class cars could be improved. This work culminated in a new, more streamlined range in 1979, featuring new engines of 3.8 and 5 litres, further extended to 5.6 litres in 1986 after the introduction of increasingly sleek small saloons. These included high-performance models with engine development assisted by the Cosworth Grand Prix firm.

SUMMARY

Plus points
- Massive performance for its day.
- Comfort.
- Mechanical durability.
- Good styling.

Minus points
- Wild handling.
- Tendency towards severe corrosion.
- High maintenance costs.
- Expensive parts.

▽ A 250 bhp 6.3-litre V8 engine gave the Mercedes 300SEL saloon a massive performance.

M.G.

The great British M.G. sports car marque was born out of one man's frustration at having to sell the mundane Bullnose Morris during the 1920s. This car and its cheaper companion, the Cowley, were sound and dependable cars, but hardly glamorous, so Cecil Kimber, manager of Morris Garages in Oxford, decided to see whether he could make his dealership more profitable by supplementing the range with more glamorous versions at a premium price. The first cars appeared in 1922 and, with their lowered suspension – like competition machines – and special paint schemes, they sold so well that by the following year they were being referred to by all and sundry as M.G.s (for Morris Garages).

The modifications soon extended to special bodies, some of which were quite extravagant, while others were very spartan, like that of a 1925 competition model which won a gold medal on an early form of sporting rally called the Land's End Trial.

Years later, when it had been rescued from a scrapyard, this car became known as 'Old Number One', simply because no one at that time could recall an earlier M.G. and because it coincided with the point at which the marque adopted the octagon as a trademark and went into serious production.

△ ***The J2 M.G. Midget was one of the most charming sports cars, starting a trend that lasted more than 20 years. Paradise Garage.***

J2

YEARS MADE: 1932–4.
ENGINE: In-line four-cylinder, 847 cc, single overhead camshaft, 36 bhp at 5500 rpm.
FRAME: Twin girders, cross-braced.
PERFORMANCE: 80 mph (129 km/h).
CURRENT PRICE GUIDELINE: * * Appreciating steadily.
PROBLEMS TO LOOK FOR: Deterioration, lack of skilled maintenance, non-originality.
CLASSIC CAR STATUS: * * * One of the most charming small pre-war sports cars.

A second distinctive M.G. feature began to emerge in the following year, 1926, when the parent Morris company abandoned its old rounded radiator in favour of an American-style flat grille which proved less aerodynamic.

As M.G.s received their own flat grille, Kimber's men began to tune the engines to make sure that no performance was lost. The process continued until the M.G. Super Sports, as it was known, became a much faster car than the basic Morris. By 1928, now called the 14/40, it was selling so well that Kimber was able to introduce a new chassis and six-cylinder twin-carburettor engine to give a completely new car, the 18/80, with a smaller four-cylinder version, the Midget, for support.

The cheap new Midget, with an advanced overhead-camshaft cylinder head, was an immediate success, accounting for half the firm's production, which itself tripled in 1929 alone. Such was the volume of cars, around 1000 a year, that the emergent M.G. Car Company had to move to Abingdon, near Oxford, in 1930.

This site, next to the old Pavlova Leather Works, was to become the world's biggest sports car factory, distinguished by its brown-and-cream colour scheme. It was also during 1930 that the enduring M.G. slogan 'Safety Fast' was adopted, and a young accountant, John Thornley, helped to start up the M.G. Car Club. He then

▷ ***The M.G. K3 was produced with various forms of bodywork. Briggs Cunningham Museum.***

joined the firm in 1931 and years later took over Kimber's mantle.

As M.G. entered the 1930s, the Midget, called the M-type, accounted for the bulk of the sales and was soon performing well on the racetrack. Because of this, special C (for competition) types were constructed. They had higher-powered engines and more aerodynamic bodywork, which enabled them to beat the factory-entered Austin Sevens in Britain's version of the Le Mans 24-hour race, the Brooklands Double Twelve.

With such a successful record, the Midgets soon became popular competition cars, winning races and hillclimbs all over the world. Kimber was not content to rest on his laurels, however, and was working hard on a new small six-cylinder model for added performance. At the same time, a highly tuned, supercharged version of the C-type Midget with ultra-streamlined bodywork, called EX120 because it was an experimental model, took a series of speed records from the Austin Seven at the Montlhéry track near Paris.

The normal 847 cc roadgoing Midget was improved in 1931 as a D-type model, with optional four-seater bodywork. Its name indicated that it was based on the C-type, rather than on the M-type which continued in production as just a two-seater. Soon afterwards an F-type, using the new 1271 cc six-cylinder engine in conjunction with the D-type's bodywork, established another line called the Magna. It all seemed very complicated, but, in reality, Kimber was simply building a wide variety of cars around two engines.

M.G.s became far more specialized in 1932 with the introduction of the J (for junior) series of Midgets, of which the J2 two-seater was the classic. It had a very high-revving, much modified engine for maximum performance, and a body with cutaway doors like those of the C-type, fold-flat windscreen and cut-off tail with Le Mans-style fuel tank. This body was to set the style for M.G. Midgets for more than 20 years. J1 versions had four seats, the J3 a supercharger, and the J4 was a racing version.

SUMMARY

Plus points

- Excellent performance by contemporary standards.
- Good handling.
- Charming appearance.
- Fascinating workmanship.
- Better-than-average availability of spares.
- Relatively economical to run.
- Numerous events in which it can compete.

Minus points

- Needs skilled maintenance by today's standards.
- Prone to engine oil leaks.
- Crankshaft trouble if overstressed.

K3

YEARS MADE: 1932–4.
ENGINE: In-line six-cylinder, 1087 cc, single overhead camshaft, supercharged, 105–125 bhp at 5500 rpm according to tune.
FRAME: Twin girders, cross-braced.
PERFORMANCE: 110 mph (177 km/h).
CURRENT PRICE GUIDELINE: **** Set to appreciate rapidly, has great rarity value.
PROBLEMS TO LOOK FOR: Non-originality, deterioration.
CLASSIC CAR STATUS: ****
Appreciating steadily.

Short-chassis versions of the Magna were developed into the K-series Magnette with revised engines, wider track, larger brakes and improved steering. Both models were available with a wide variety of bodywork, with the ultimate K3 supercharged Magnette being produced only in small quantities.

The K3 achieved numerous successes in competition, including winning the team award in the 1933 Mille Miglia. Also, the greatest racing driver of the day, the Italian Tazio Nuvolari, drove a K3 to victory in Britain's oldest road race, the Tourist Trophy, in a time that was not bettered until Stirling Moss won the race with a Jaguar XK120 in 1951! A highly aerodynamic single-seater K3, called EX135, also established numerous speed records over a similar period.

SUMMARY

Plus points

- Rarity value.
- Excellent performance and handling.
- Relatively durable for pre-war supercharged racing car.
- Fascinating workmanship.
- Good appearance on average.
- Numerous events in which it can compete.

Minus points

- Needs skilled maintenance.
- Expensive to restore.

TC

YEARS MADE: 1946–9.
ENGINE: In-line four-cylinder, 1250 cc, overhead valves, pushrods, 54 bhp at 5200 rpm.
FRAME: Twin girders, cross-braced.
PERFORMANCE: 78 mph (126 km/h).
CURRENT PRICE GUIDELINE: ** to ***
According to condition.
PROBLEMS TO LOOK FOR: Deterioration through age.
CLASSIC CAR STATUS: ***
Appreciating.

By the early 1930s, technological development was running riot at Abingdon, with all manner of Midgets, Magnas and Magnettes being produced in open, fixed-head coupé, or saloon form, and with two- or four-seater bodies, until it seemed as though hardly any two cars were the same. However, the next really outstanding model was the PA Midget of 1934 which had a more advanced chassis, similar to that of the N-type Magnette, the same powerful brakes, and a stronger engine with three main bearings instead of two. These cars, usually painted in the works colours of brown and cream – although they were all supposed to be private entries – became star attractions in the reliability trials which were the predecessors of our modern rallies.

Work was progressing on an all-independently sprung M.G. before the parent Morris company, under the founder William Morris, and his chief executive Leonard Lord (later to move on to Austin and the British Motor Corporation), intervened. They called a halt to M.G.'s never-ending developments, which were using up all the profits. The workforce was, quite naturally, horrified, but work was still scarce in post-Depression England and they had to knuckle down to building simpler, more profitable cars.

The first of these was the TA Midget, which replaced all the previous small sports cars. The components came mostly from the basic Morris Ten saloon and, as such, it had a much more civilized ride than its very stiffly sprung predecessors could offer. Its pushrod engine was also cheaper and less efficient, but the performance of the new car did not suffer because the capacity, at 1292 cc, was larger. Much to Abingdon's surprise, sales did not suffer either! This was partly because the TA was more flexible, and easier to drive as a result, partly because it had new hydraulic brakes, partly because Kimber's styling was as good as ever, and partly because the new M.G. still offered the best value for money.

The Midget was M.G.'s mainstay in the late 1930s, entering the war years as a TB with a revised engine so that it was able to emerge in 1945 with minimal modifications as the TC. It was this model, more than any other, which pioneered the boom in British car exports. Despite the fact that the TC was beginning to look antiquated, sales tripled, simply because there were ample supplies of the car.

▽ ***The M.G. TC became the ultimate – and highly popular – extension of the J2 theme. Owners: Naylor Brothers.***

Abingdon was able to get the steel it needed to produce all these TCs because a large proportion of the cars were exported.

The TC was updated as the TD in 1949, with independent front suspension from the similar Y-type M.G. saloon. Meanwhile, Abingdon worked on a new version with modern full-width bodywork. However, despite the efforts of their new general manager, John Thornley, the design was 'frozen' when Lord took over the British Motor Corporation, formed from the merger of Austin and Morris's interests. The Austin-Healey of 1952 was the result (see page 28), with Abingdon being forced to soldier on until 1955 with the TF face-lifted version of the Midget.

SUMMARY

Plus points

Charming appearance.

Forgiving handling.

Decent performance.

Very durable if well maintained.

Good availability of spares.

Numerous events in which it can compete.

Minus points

Very hard ride for a post-war car.

Needs higher standard of maintenance than is common nowadays.

MGA TWIN CAM

YEARS MADE: 1958–60.
ENGINE: In-line four-cylinder, 1588 cc, twin overhead camshafts, 180 bhp at 6700 rpm.
FRAME: Twin girders, cross-braced.
PERFORMANCE: 117 mph (188 km/h).
CURRENT PRICE GUIDELINE: * * Set to appreciate.
PROBLEMS TO LOOK FOR: Engine can be delicate, corrosion may be a problem in chassis and body.
CLASSIC CAR STATUS: * * Appreciating gradually.

When the Abingdon team was at last allowed to introduce the MGA, it became a best-seller. This key car was available in either open or, alternatively, Jaguar XK120-like fixed-head form with, initially, a 1500 cc engine which provided a 100 mph (161 km/h) performance. It is interesting to note that the bodywork was based on the shape of the pre-war EX135 project, which had still been continuing to set new records into the 1950s, and on the similar EX179.

The basic pushrod MGA, which was uprated first to 1588 cc and eventually to 1622 cc, continued to sell well until 1962. Once the MGA was on the stocks, however, M.G. designer Syd Enever promptly started building a new project car, EX181. The contemporary M.G. saloons matched Lord's ideals better in that they were simply re-badged versions of BMC base models, rather than separate designs.

Experiments with two twin-cam cylinder head conversions – one of which was H.R.G.'s (see page 80) – had resulted in 1958 in a high-performance 1588 cc version of the MGA, aimed specifically at production sports car racing. This car had disc brakes all round and a top speed of 117 mph (188 km/h). Unfortunately it suffered from oil consumption problems and had to be withdrawn in 1960 . . . after which, ironically enough, the problems were eventually cured!

SUMMARY

Plus points

Excellent performance and handling.

Relatively easy to rebuild.

Numerous events in which it can compete.

Fixed-head version very pretty.

Minus points

Engine needs careful maintenance, but durability much improved now.

Tendency towards corrosion in chassis and body.

Wheels can be overstressed.

▽ ***The standard MGA became the high-performance Twin Cam for competition.***

△ ***The MGB, pictured in its final, rubber-bumpered form. Performance Cars Ltd.***

MGB

YEARS MADE: 1962–80.
ENGINE: In-line four-cylinder, 1798 cc, overhead valves, pushrods, typically 95 bhp at 5400 rpm.
FRAME: Integral.
PERFORMANCE: 106 mph (171 km/h).
CURRENT PRICE GUIDELINE: * to ** Depending on condition and specification.
PROBLEMS TO LOOK FOR: Corrosion.
CLASSIC CAR STATUS: * Set to appreciate.

The Austin-Healey 'Frogeye' Sprite (see page 29) was restyled along more conventional lines to be produced in large quantities with a 1098 cc engine as a new Midget from 1961. The styling theme was based on EX181's aerodynamic shape and bore a distinct resemblance to what would be the most popular M.G. of all, the MGB.

More than 500,000 1800 cc MGBs would be made between 1962 and 1980 to fill the need for a cheap sporting vehicle that was solid, reliable, and available in both open and hardtop forms. Initially, it was a two-seater, but from 1965 the MGB was available as the fixed-head GT with two extra tiny rear seats. In this form, the MGB GT featured a hatchback door like that pioneered by Aston Martin's DB Mark III to accommodate far more luggage than was possible in the roadster.

During 1967 the MGB was developed into the MGC to replace the ageing Austin-Healey 3000 (see page 30), but Austin's new engine proved to be so heavy that the handling was compromised, and what should have provided strong opposition for Datsun's 240Z (see page 56) lasted only two years.

Only minor changes were made to update the MGB and Midget – except for an exciting Rover V8 engine for the BGT (killed by the world oil crisis) and a larger Triumph engine for the Midget from 1974. This last was because the British Motor Corporation was trying to deal with the desperate need for a new Triumph sports car. Eventually the MGs became so outdated that Abingdon was closed down, partly due to a currency crisis in their largest market, America.

The MG name continued, however, on upmarket editions of Austin's small Metro saloon. Then a pure competition model, the Metro 6R4 supercar, with a 3-litre racing engine and spaceframe, was built from 1983 for the short-lived Group B international series.

SUMMARY

Plus points
- Very durable mechanically.
- Excellent handling.
- Reasonable performance.
- Good appearance.
- Numerous events in which it can compete.
- Excellent availability of reasonably priced spares.

Minus points
- Tendency towards terminal corrosion in more attractive chrome-fronted earlier models.
- Handling not so good on later rubber-bumper cars.

MINI

Few cars, let alone one so small, have made such an impact on the world as the Mini. Conceived during the world's first oil crisis in 1956, the Mini was the last great product of Leonard Lord. He realized that there was a huge demand for a very economical car like the old Austin Seven (see page 26); one that was superior to contemporary 'bubble cars', which were seen as being too much like motor cycles. It was also the greatest design of Alex Issigonis, who already had the brilliant Morris Minor to his credit.

MINI

YEARS MADE: 1959–67.
ENGINE: In-line four-cylinder, 848 cc, overhead valves, pushrods, 34 bhp at 5500 rpm.
FRAME: Integral.
PERFORMANCE: 72 mph (116 km/h).
CURRENT PRICE GUIDELINE: * Will appreciate when present cars are out of production.
PROBLEMS TO LOOK FOR: Corrosion, neglect.
CLASSIC CAR STATUS: *** Set to appreciate.

△ ***Rarer than rare . . . one of the first Mini saloons, produced in 1959, that changed the face of motoring.***

Towards the end of the 1950s, the average Briton's ambitions were modest: small meant secure and affordable and this attitude spilled over into car fashions, thereby benefiting the Mini. Mini sales were helped by the 'swinging sixties' too, when 'dedicated followers of fashion' decided that it was chic and bought them by the thousand. Outside Britain, it sold well everywhere except in the United States, where typically long journeys and cheap fuel were incompatible with the economical little car that many thought too cramped for comfort.

With the exception of the Americans, few people minded the Mini's small passenger compartment – you could still get four people inside this bargain-priced box – nor did they object if the engine was already growing old, gears howled, or wind whistled around ill-secured windows. They did not mind the spartan interior, the sit-up-and-beg driving position, and the windscreen wipers which you had to park by yourself as you mopped away misty patches on the glass. Trendy young people looking for a car they could park in bustling city centres had shown that these incredibly cheap and economical cars could transcend all class barriers. Racing drivers had shown that the Mini's amazingly stable handling could turn it into a giantkiller. In no time at all, the love affair turned into idolatry: Minis were adorned with all manner of accessories – decorations which demonstrated that, although the owner was one of the people, he or she was really an individual.

Issigonis's greatest feat was squeezing so much into what was, in reality, a 10-ft (305-cm) square tin box. He did it by turning the engine round to run transversely across the car, driving the front wheels, while further space was saved by fitting the transmission in the sump. This was made possible by pioneering achievements in the economical manufacture of constant velocity joints for the shafts which not only had to drive the wheels, but also allow them to be steered as well. Mounting the gearbox in the engine's sump so that it had to share the same lubrication was made possible by dramatic improvements in oil. A third factor which enabled Issigonis to build what is still the smallest comfortable four-seater was the development of tiny, but reasonably durable 10-in (25-cm) diameter tyres. In addition, the rough ride which would normally have been expected with such small wheels was largely countered by revolutionary rubber suspension, and this also saved a lot of space. The combination of front-wheel drive (which avoided expense and the encroaching bulk of a propellor shaft), the need to have a wheel at each corner, space-saving rack-and-pinion steering and a very low overall build, gave the Mini exceptionally good handling.

It was so much lighter than most of its contemporaries when it was introduced in 1959 that its makers, the British Motor Corporation, reduced the size of the engine from 948 cc to 848 cc to keep its speed down, doing away with the need for more expensive brakes and avoiding a higher insurance premium.

SUMMARY

Plus points

- Excellent performance.
- Good handling.
- Durable mechanical parts.
- Easy to restore.
- Major parts readily available.

Minus points

- Corrosion.
- Interior wears quickly.
- Hydrolastic suspension parts obsolete.

MINI-COOPER S

YEARS MADE: 1963–71.
ENGINE: In-line four-cylinder, 970 cc, 1071 cc or 1275 cc, overhead valves, pushrods, 65 bhp at 6500 rpm, 70 bhp at 6200 rpm or 76 bhp at 5800 rpm.
FRAME: Integral.
PERFORMANCE: 93 mph (150 km/h), 94.5 mph (152 km/h), or 96 mph (154 km/h).
CURRENT PRICE GUIDELINE: ** Set to appreciate rapidly.
PROBLEMS TO LOOK FOR: Corrosion, neglect, mistreatment, non-originality.
CLASSIC CAR STATUS: *** Set to appreciate.

Not long after the Mini's introduction, a whole new industry mushroomed, selling tuning kits to make standard Minis go faster. Eventually, world champion racing car constructor John Cooper persuaded BMC to market a faster version with better brakes called the Mini-Cooper, which appeared in 1961.

This exercise was such a success that even quicker versions, called the Mini-Cooper S, were introduced in 1963, and they gave factory-entered competition cars numerous victories, including a hat trick in the Monte Carlo Rally. Attempts to standardize the BMC range on a fluid-based Hydrolastic suspension system from 1964 did not work so well with the short-wheelbase Mini, however, and it reverted to the rubber system from 1970.

A wide range of Mini-based vehicles, including vans, estate cars and small open trucks called pick-ups, were marketed until 1981, when British Leyland, as BMC had become known, introduced a larger, more modern small car called the Metro. But the Mini saloon, with ever more sophisticated trim, remains in production because there seems to be an insatiable demand for the smallest full-sized car in the world.

SUMMARY

Plus points

- Compact.
- Superb handling.
- Durable mechanical parts.
- Very economical.
- Easy to restore.
- Vast majority of parts readily available.

Minus points

- Corrosion.
- Interior wears quickly.

▽ The Mini-Cooper S thrilled spectators everywhere in the thick of competition.

MORGAN

Morgan has become the last survivor of what was once a popular machine . . . the traditional open sports car. As the modern car industry's mass-produced models bristle with more electronic gadgetry and cosseting luxuries each year, Morgan offers a way of reliving the 'good old days' of open-air motoring.

4-4

YEARS MADE: 1936–9.
ENGINE: In-line four-cylinder Coventry Climax, 1122 cc, inlet valves over exhaust, 34 bhp at 4500 rpm.
FRAME: Twin girders, cross-braced.
PERFORMANCE: 75 mph (121 km/h).
CURRENT PRICE GUIDELINE: *** Will appreciate gradually.
PROBLEMS TO LOOK FOR: Non-originality, deterioration through age.
CLASSIC CAR STATUS: ** Will appreciate.

The 4-4's design was conservative in many aspects, even when it was first introduced in four-wheel form in 1936: all previous Morgans had been three-wheelers powered by motor cycle-type engines. These three-wheelers were unusually advanced when they were launched in 1910 in that they had independent front suspension!

This system, based around two vertical pillars with sliding links to the axles on which the front wheels were mounted, was based on motor-cycle practice and was later to be adopted by Lancia with its history-making Lambda (see page 100). One of the main advantages of Morgan's suspension was that it enabled the steered wheels to stay in contact with the road surface at almost all times. This meant that a Morgan's steering was always much better than that of typical contemporary vehicles which relied on a car-style solid front axle bouncing from bump to bump so that the wheels spent little time in contact with what were frequently very rough roads.

Naturally, the three-wheel Morgans soon started showing well in competition, especially because their chassis could handle more power than many others due to their front suspension. However, with the advent of the Austin Seven (see page 26), the days of three-wheelers – which cost about the same but offered far less accommodation and stability – were numbered, and eventually even the conservative Morgan company had to introduce a four-wheeler.

It was only natural for this model, called the 4-4 (for four wheels and four cylinders, as many previous Morgans had been powered by twin-cylinder motor-cycle engines), to be based on the existing three-wheeler practice. In engineering terms it amounted to little more than an extended chassis frame with two wheels at the back instead of one, and a more powerful and flexible four-cylinder Coventry Climax engine to haul the extra weight. However, although the 4-4 had conventional lines for its day – very much like the M.G. Midget's (see page 125) – they were exceptionally attractive, and the car was soon available in four-seater form to supplement the normal two-seater range. Morgan was also able to offer a more luxurious drop-head coupé version without disruption to the normal production of around four cars a week because the company was in the unusual position of building its own bodywork and not having to rely on outside sources.

SUMMARY

Plus points

- Oozes charm.
- Good appearance and handling.
- Reasonably spritely performance.
- Easy to maintain.
- Economical to run.
- Numerous events in which it can compete.

Minus points

- Engine can give problems.
- Corrosion often present.

▽ ***The oldest surviving Morgan 4-4 in original condition, by courtesy of John Orton.***

PLUS FOUR SUPER SPORT

YEARS MADE: 1961–6.
ENGINE: In-line four-cylinder, 2138 cc, overhead valves, pushrods, 115 bhp at 5000 rpm.
FRAME: Twin girders, cross-braced.
PERFORMANCE: 110 mph (177 km/h).
CURRENT PRICE GUIDELINE: * * * Set to appreciate.
PROBLEMS TO LOOK FOR: Corrosion, misuse.
CLASSIC CAR STATUS: * * Underrated.

When supplies of the small Coventry Climax engine dried up in 1939, Morgan switched to a similar unit produced specially for the company by Standard, who continued the association after the war. Then, when the small Standard engine became uneconomical to make, Morgan revamped the 4-4's design in 1950 to take the much more powerful Standard Vanguard unit. The result was the Plus Four.

The strength of the family firm that owned Morgan was never more evident than when it managed to resist a take-over bid by Standard which was looking for a short cut to marketing a new Triumph sports car in 1952 to compete with the Austin-Healey 100 (see page 28). Morgan kept its engine supplies because its products had become such a dearly loved part of the English motoring scene.

By this time, M.G. was the only other make still producing a traditional sports car in any quantity and, once the M.G. TF had been replaced by the MGA in 1955 (see page 128), Morgan was left on its own.

As a second line, the company reintroduced the smaller, and cheaper car, now called the 4/4, with a Ford Anglia engine. Meanwhile, the Plus Four carried on using Standard's big four-cylinder power unit from the Triumph TR sports cars.

A brief flirtation in 1963 with an alternative, all-enveloping, modern fixed-head, glass-fibre body on the same chassis – marketed as the Plus Four Plus – was not well received, so it was discontinued.

The Plus Four remained the top model, with the constantly uprated 4/4 still being powered by Ford engines of ever-increasing power and capacity.

During the late 1950s, Morgan's racing reputation was revived by the activities of tuner Chris Lawrence, whose more powerful Le Mans conversion was marketed as a new model, the Plus Four Super Sport, from 1961.

△ Morgan's Plus Four Super Sport turned out to be a surprisingly capable performer in competition.

Lawrence then took his personal, modified Plus Four Super Sport, registered TOK258, with co-driver Richard Shepherd-Barron, to the marque's greatest performance: 13th place at Le Mans in 1962 against far more modern outright sports racing cars.

Normal Plus Fours continued to be sold alongside the new model until the Triumph TR range went over to longer six-cylinder engines in 1967.

In 1982, the Plus Four was reintroduced with a Fiat twin-cam engine of 2-litre capacity.

SUMMARY

Plus points
- Excellent performance.
- Good appearance and handling except for bump steer.
- Durable mechanical parts.
- Numerous events in which it can compete.

Minus points
- Very hard ride.
- Chassis and body corrodes.

PLUS EIGHT

YEARS MADE: 1967 to date.
ENGINE: V8, 3528 cc, overhead valves, pushrods, typically 115–200 bhp at 5280 rpm.
FRAME: Twin girders, cross-braced.
PERFORMANCE: Typically 120 mph (198 km/h).
CURRENT PRICE GUIDELINE: ** to *** According to age.
PROBLEMS TO LOOK FOR: Corrosion in older versions.
CLASSIC CAR STATUS: ** Set to appreciate.

After the Triumph TR range went over to longer six-cylinder engines, in 1967 Morgan once more managed to revamp the top model as a Plus Eight with a Rover V8 engine to produce the fastest Morgan yet. Because of its light weight, it proved capable of exceptionally fast acceleration. On the debit side, if offered a distinctly vintage, very stiff, ride.

Morgan was still enjoying long waiting lists well into the 1980s with cars which were, in essence, very much like the pre-war machines. They still relied on the original design of front suspension; but it had stood the test of time so well and was proving suitable for the latest developments in low-profile sports tyres!

SUMMARY

Plus points
- Excellent performance.
- Good handling and appearance except for bump steer.
- Durable mechanical parts.
- Numerous events in which it can compete.

Minus points
- Very hard ride.
- Chassis and body corrodes without careful maintenance.

▷ ***The insertion of the Rover V8 engine in the lightweight Morgan chassis produced the very fast Plus Eight.***

MORRIS

Cycle maker William Morris laid the foundations of the British motor industry with his great rival, Herbert Austin. His company was to produce – thanks to the talents of one of the world's greatest designers – a historic and evergreen model: the post-war Morris Minor.

MORRIS MINOR

YEARS MADE: 1948–71.
ENGINE: In-line four-cylinder, overhead valves, pushrods, typically 1098 cc, 48 bhp at 5100 rpm.
FRAME: Integral.
PERFORMANCE: 73 mph (117 km/h).
CURRENT PRICE GUIDELINE: * Set to appreciate gradually.
PROBLEMS TO LOOK FOR: Corrosion, misuse, neglect.
CLASSIC CAR STATUS: ** Underrated.

Morris's first car – the famous 'bullnose' of 1913 – helped turn his factory at Cowley, near Oxford, into one of the most productive in the world, along with Austin's headquarters at Longbridge, in Birmingham. Then, as Austin floundered, to be saved by the Seven (see page 26), Morris forged ahead with his medium-sized Oxford and the cheaper Cowley (on which the more specialized M.G.s were based) until, by 1925, annual output was up to 54,000.

Meanwhile, the old-established Wolseley company had run into trouble during the economic depression of the 1920s and in 1926 was eventually bought by Morris, who was flush with profits. Wolseley had a brilliant new single-overhead-camshaft engine which in 1928 found its way into the first Morris Minor, a small car which went on to pose the strongest threat to the Austin Seven and, incidentally, formed the basis of the M.G. Midgets (see page 125).

Much of Morris's success had been based on the outcome of cut-throat price-slashing campaigns which bit deep into his profits when the American-owned Ford company made it a three-way contest with Austin from 1932. In 1934, Morris hit back, however, with a new small saloon, the Eight, inspired by Ford's new Model Y. This, like the Seven for Austin, more or less saved his company. Some of the engineering, such as a side-valve power unit, was hardly inspiring, but it was cheap and durable. Much of the credit for the success of the Eight went to former Wolseley engineer, Leonard Lord, but Morris was a difficult man to work with, so the equally strong-willed Lord left for Austin in 1938.

Nevertheless, Morris again came good that year with the Series M Ten, of advanced new unitary construction like

◁ ***The Morris Minor – Alex Issigonis's first great design – was 1948's most advanced small car.***

that of the previous year's Vauxhall – built by General Motors' British subsidiary. Meanwhile, the other big British manufacturer, the Rootes Group, whose cars included Hillmans and Humbers, had some extraordinarily talented engineers, including William Heynes – who left in 1935 to become Jaguar's chief engineer – and a young Briton of Greek and German extraction called Alex Issigonis.

Issigonis joined Morris in 1936 to further his career and, after contact with Hubert Charles, who had been responsible for M.G.'s ill-fated all-independently sprung R type racing car in 1934, designed an exciting new wishbone and coil system of independent front suspension for the Series M Ten. Part of the concept included equally brilliant rack-and-pinion steering. Morris's hierarchy considered such technology too advanced, however, so the Series M Ten was introduced with a conventional beam front axle, although it did have unusually flexible leaf springs which gave it a good ride.

The new steering and suspension from Issigonis made its début on the small M.G. Y-type saloon in 1947. However, Issigonis had received powerful backing from Morris's managing director, Miles Thomas, to design a new small Morris for post-war production. In fact he had been given an almost unique opportunity to work entirely to his own ideas from a clean sheet of paper, at a time when most mass-produced cars were already being designed 'by committee' around existing components.

Issigonis responded with what, in 1948, would be the most advanced small car in the world – the Morris Minor. Its most outstanding feature was that everything in the design was extraordinarily well integrated, fully justifying Thomas's brave decision to entrust the work to one man. Under the circumstances, it was ironic that Thomas had had to resign because of the usual difficulties in dealing with the ageing Morris, now Lord Nuffield.

One of the most adventurous aspects of the Morris Minor was its use of what were, for those days, extremely small wheels. Issigonis pointed out that they were the best way to lower the car's centre of gravity, reduce its unsprung weight, and save space generally. Naturally the design also used independent front suspension by space-saving wishbones, and rack-and-pinion steering, which combined with an engine mounted far forward, and the other features, to provide exceptionally good and stable handling.

But the most striking feature of the Morris Minor, which stayed in production for 23 years (when many cars last only seven), was its appearance: it looked like no other car, because it had been designed entirely functionally, and by one man – Issigonis. His genius was to be seen yet again in his next great work, the Mini (see page 130).

SUMMARY

Plus points

- Dependable.
- Excellent handling.
- Great character.
- Economical.
- Ready availability of spares.
- Easy to maintain and restore.

Minus points

- Tendency towards corrosion when neglected.

NSU

The renowned German motor cycle manufacturer, NSU, threw caution to the winds when it designed its greatest car, the Ro80 . . . and it killed the company. An even greater tragedy was that the car promised so much: for the first year or so from its introduction in 1967 there was nothing on the roads which could match it. No other saloon car could offer such a marvellous combination of comfort, supple handling, beautifully weighted steering and braking, and supremely elegant aerodynamics with a unique engine. Jaguar would match the NSU with its XJ6 a year later (see page 89), but that had far lighter steering than almost any European liked. The glaring difference, however, lay in the engines: Jaguar's was the supremely smooth and powerful in-line six-cylinder XK unit, while NSU's was the one the world expected to take over the industry – Felix Wankel's rotary engine.

Ro80

YEARS MADE: 1967–77.
ENGINE: Twin rotor Wankel, 994 cc, single inlet and exhaust port per rotor, 114 bhp at 5500 rpm.
FRAME: Integral.
PERFORMANCE: 107 mph (172 km/h).
CURRENT PRICE GUIDELINE: * Will appreciate dramatically when the world has forgotten the engine's reputation.
PROBLEMS TO LOOK FOR: Engine wear, corrosion.
CLASSIC CAR STATUS: **** Years ahead of its time.

NSU began planning what it expected to be a world-beater in 1961. By 1963 it had been decided that it would have the sensational new Wankel engine, linked to semi-automatic transmission with front-wheel drive and very aerodynamic styling. The engine, in particular, in which a spinning rotor replaced both pistons and valves, meant breaking new ground to such an extent that it was worth putting a small car into limited production to see how it stood up to normal use over a prolonged period, before going into full-scale production.

△ **The NSU Ro80 promised to be great, but was let down by its Wankel engine. Owner: Obert.**

The result was the tiny Bertone Spider of 1964, with a single-rotor Wankel engine of only 500 cc capacity. (The displacement is normally doubled for comparison with conventional engines.) This Spider was the world's first production rotary-engined car, introduced just before one from NSU-licensee Mazda in Japan. With hindsight, it should have been NSU's last such car without a dramatic rethink. The Bertone Spider produced enough power for a good performance, but suffered from a very heavy fuel consumption, not much pulling power, and a dreadful lack of reliability from its overstressed rotor tips.

NSU convinced itself, though, that all the engine needed was a little development to improve its durability . . . and fuel was far less expensive in the 1960s. The company reasoned that few people would complain at a high consumption in return for an engine which was not only unique but, quite literally, as smooth as a turbine.

So NSU shut its eyes to reality and went ahead with a double-size Wankel engine with twin rotors in the hope that it would be less highly stressed and would not overheat and begin to break up like the smaller one had done. The rest of the new car was superb, so NSU must strike while the iron was hot!

In retrospect, the Ro80 (to convey a rotary which would take NSU into the 1980s) can be seen as the forerunner of a whole new breed of highly aerodynamic cars which, largely speaking, did not appear until the 1980s. The factor by which such a windcheating ability is measured is called the Cd (coefficient of drag), and the Ro80's, at 0.35, is only now being equalled and overtaken. What could not be measured was the superb styling that went with these lines – attributed to Italian Pio Manzu.

On the road, the Ro80 was supremely stable at speed, even if its engine's enormous appetite for fuel (12 mpg, or 24 l/100 km, was not uncommon) offset such advantages. The Ro80 was also extremely well made, and rather heavy as a result, which did little to help its fuel consumption and meant that the engine had to labour to produce reasonable acceleration. However, if you were willing to drive the engine hard, the Ro80 was quite spritely, because it had a servo-assisted clutch linked to its wide-ratio three-speed gearbox.

The coil spring suspension and disc brakes on all four wheels – with a history-making inboard siting at the front to reduce unsprung weight – were superb. However, they were not enough to make up for engines which frequently failed to outlive their guarantee period and cost NSU – and its ultimate owners, Volkswagen and Audi

– a fortune. Within a couple of years, the Ro80 had acquired a deserved reputation for being a superb car with a terrible engine.

Fundamental changes in the design and materials developed in Britain by Rolls-Royce improved the Wankel greatly, and it eventually made an excellent engine for Mazda. Sadly, these developments came too late to save the Ro80's reputation, which plunged even lower after the world's first energy crisis in 1973. NSU lingered on with the Ro80 until 1977 before, tragically, Volkswagen and Audi had to bring down the axe. The ghost of the Ro80 lives on, however, in the shapes of their cars today.

SUMMARY

Plus points

- Superb appearance and aerodynamics.
- Beautiful steering and brakes.
- Supple suspension.
- Effortless high-speed cruising.
- High standard of comfort.

Minus points

- Rotary engine smooth but disastrously underdeveloped with major durability problems.
- Inevitable corrosion.

OLDSMOBILE

'It goes beyond elegant looks and rich interiors. Beyond boulevard ride. To the ultimate luxury in action: front-wheel driving. Men admire its command of the road. Women, its obedience. It tracks where others can't. Smoothly. Silently. Toronado's deeper luxury shows. In surprising room. Flat floors. New GM safety features. Let the youngmobile thinking in luxury spoil you now – at your nearest Olds dealer. Escape from the ordinary.'

So exhorted General Motors' advertising for its great technological breakthrough of the 1960s, the massively reliable Oldsmobile Toronado.

TORONADO

YEARS MADE: 1965–70.
ENGINE: V8, overhead valves, pushrods, typically 7456 cc, 375 bhp at 4600 rpm.
FRAME: Integral.
PERFORMANCE: 131 mph (211 km/h).
CURRENT PRICE GUIDELINE: ** Set to appreciate.
PROBLEMS TO LOOK FOR: Corrosion.
CLASSIC CAR STATUS: ** Technical leader.

General Motors had made the medium-priced Oldsmobile brand, one of the oldest motor manufacturers in America, its technical leader from the 1920s onwards. With Oldsmobile models it introduced brakes on all four wheels in 1927, synchromesh gearboxes in 1931, a new straight-eight-cylinder engine in 1932, independent front suspension in 1934, controversial 'turret top' styling in 1935, and the first GM automatic gearbox in 1938, which had developed into the superb Hydramatic by 1940. A 'Futuramic' styling revolution with such noteworthy features as a split curved windscreen and wraparound rear window was launched in 1948, followed in the next year by a new power unit called the Rocket. This Oldsmobile V8 engine was the first of its kind with modern 'oversquare' dimensions, and after it came the all-alloy lightweight V8 shared with Buick in 'compact cars' from 1960. By 1963 the Jetfire line had a turbocharger for added performance . . . but the biggest breakthrough came in 1965 with the Toronado.

The six-seater two-door coupé not only had front-wheel drive, but it was also the first car of its type to combine

▽ The Oldsmobile Toronado represented General Motors' most advanced technological thinking. Owner: Earl Orm.

that with automatic transmission and a really powerful engine – Oldsmobile's biggest. The result was a car with a vast amount of room inside (because it had no transmission tunnel), ease of access and exit through the biggest doors of any car in America (they measured more than 3 ft across) and a supremely smooth delivery of power through the famous Hydramatic transmission from Oldsmobile's giant 7-litre (later 7.5-litre) V8 engine. It was also one of the best-looking cars General Motors had designed in years, setting a clean new trend in styling.

The Toronado was no sudden last-minute effort inspired by Alex Issigonis's sensational front-wheel-drive Mini (see page 130). General Motors had been working on the design since 1954 when it set up a development group to produce a space-saving Unitized Power Package (UPP for short) to combine the engine, transmission and final drive in one unit, as Issigonis had done. The concept was first seen in an absolutely ghastly-looking La Salle show car in 1955. By 1958, the design group had the UPP fully functional, using a 7-litre V8, General Motors' largest engine at the time, which drove the front wheels because it was too heavy to be carried in the back of a car. In normal applications, little more than 150 bhp could be transmitted comfortably through wheels which also had to be steered, but this problem was overcome so well with power-assisted steering that it proved possible to convey as much as 400 bhp.

General Motors then looked for a market for its marvellous new invention and chose the Oldsmobile customers. This was after a survey in 1960 showed that potential owners of a luxurious sporting car were the ones who would be most receptive to a model which boasted features that had not been offered before. By 1961 it had been decided that – in order to maximize profits – this should be what Americans knew as a 'full-sized' car and the rest of the world would call a giant.

A design competition was then instigated among General Motors' assembled stylists. It was won by assistant chief designer David North with a full-sized model called 'the Red Rendering' because it was in clay. It was only then, after several months' consideration, that General Motors' management confirmed that North's design would be combined with the UPP to form the new Oldsmobile. By the end of 1963, a running prototype was ready for managerial evaluation, and feverish work would be needed to get it into full-scale production by 1965. It was only at that point, after years of collective head-scratching, that it was decided to use a redundant Chevrolet show car name for the new machine: Toronado.

For the first few months, General Motors could not build enough of them. A spokesman said: 'It seemed there was a guy on every block who wanted to be the first to own one. Then all of a sudden the volume dropped.' Oldsmobile almost ran out of customers who wanted to be daring aesthetically and mechanically. The bottom had virtually dropped out of Toronado sales by 1967, only for them to be revived by an even bigger, 7.5-litre engine, followed by detail styling changes that eventually resulted in a new, conventional body from 1970.

SUMMARY

Plus points

- Massive reliability.
- Technological leader.
- Good appearance of original models.
- Better than average traction.

Minus points

- Tendency towards corrosion in damp climates.
- Brakes of early models poor by modern standards.

PACKARD

The golden era of Packard, a marque once called the Rolls-Royce of America because its cars and aero-engines were so well made, started in 1915 when it introduced the world's first production V12 engine, based on that of a competition car built by the British firm Sunbeam. The V12, or 'Twin Six' concept, was to form the backbone of Packard's top-of-the-range production for most of the marque's existence.

PACKARD TWIN SIX

YEARS MADE: 1931–9.
ENGINE: V12, 7297 cc, side valves, 150 bhp at 2800 rpm.
FRAME: Twin girders, cross-braced.
PERFORMANCE: 93 mph (150 km/h).
CURRENT PRICE GUIDELINE: * * * * Still appreciating.
PROBLEMS TO LOOK FOR: Sheer old age.
CLASSIC CAR STATUS: * * * * Monument to a bygone age.

Packard's early Twin Six cars, as they were called, were beautifully made, but they were very expensive. A bare chassis, for instance, cost more than a complete Cadillac of the most opulent specification from rivals General Motors. The early power unit also inspired the famous Liberty aero-engines built by Cadillac's Henry Martyn Leland during the First World War.

After the war, these splendidly smooth-running Packard cars became the traditional transport of American presidents. Warren G. Harding showed the way with a 1921 model, preferring it to a horse-drawn carriage and thereby staging the first motorcade to the White House. Later, bulletproof Twin Sixes were especially popular with Franklin D. Roosevelt, and elements of their construction were used in US Secret Service vehicles, and those for such notable personalities as Prohibition runner Al Capone.

It was a case of the more cylinders the merrier for Packard, except in 1922 when it decided to out-manoeuvre General Motors with a straight-eight. Packard's superb self-taught designer, Colonel Jesse G. Vincent, produced a magnificent vehicle which took Packard to the top of the American luxury car market.

In the late 1920s, some customers still hankered after a V12 and, when Cadillac came out with its V16, Packard felt that it had little option but to recreate the legendary Twin Six. However, the management realized that the market was limited during the days of the Depression, and consequently decided to stand apart from the rest of the American auto industry by resisting potentially ruinous annual model changes. Thus, the first new Packards built in 1931 were called the Ninth Series, harping back to a glorious past, between 1915 and 1922, when the Twin Six had been changed every year. The new Twin Six was an especially significant car in that, although it was to be produced only in small numbers, it had many new features which were to appear on lesser Packards of the Ninth Series, and these were going to sell by the thousand.

▽ The Packard Twin Six pioneered thc marque's V-shaped radiator that became a trademark. J. Dougherty Ltd.

The Twin Six pioneered Packard's famous V-shaped radiator and the movement of massive spring-loaded bumpers was damped by oil to iron out unwanted chassis flexing. The braking system for this heavy car was boosted by a vacuum servo, while the aluminium-headed engine was given dual downdraught carburettors.

Initially, the new Twin Six was available only with coachwork by Dietrich to four styles: a Sport Phaeton, a Fixed-head Coupé, a Convertible Coupé and a Convertible Sedan – all vastly overpriced. Demand for these cars was so sluggish that the 1932 production run did not sell out until 1934, proving the wisdom of avoiding annual model changes. This also meant, of course, that the company did not have to keep any particular model in production for a whole year, but could introduce a new series whenever the time seemed right. Modifications were brought in over the years, and the Twin Six went on to a Seventeenth Series before bowing out for the Second World War.

After the war, Cadillac's adventurous styling left the conservative Packard Motor Company stranded with its traditional upright radiator grille as it did not wish to be seen copying the competition. The cars were now based on a lower-priced straight-eight-cylinder model which had been introduced in 1937, or a smaller straight-six. Although the old engines were starting to look very dated by the 1950s, the by-then underfinanced Packard did manage to introduce automatic transmission in 1949, and power-assisted brakes, steering and window lifts by 1954. A merger with Studebaker in that year brought a much-needed new overhead-valve V8 engine, but continually falling sales led to the marque's extinction in 1958.

SUMMARY

Plus points

Magnificent construction.

Superb traditional appearance.

Exceptional durability.

Minus points

Very high costs of restoration.

PEGASO

One of the strangest and most exotic cars ever built was the Spanish Pegaso: strange, especially, because it was totally unlike anything else made in a country which had just been ravaged by civil war. The Pegaso name was initially intended to be synonymous with a world-beating national truck. Happily, it left the world a more memorable legacy.

Z102B

YEARS MADE: 1953–5.
ENGINE: V8, 2816 cc, twin overhead camshafts, 185 bhp at 6300 rpm, or 230 bhp at 6000 rpm with supercharger.
FRAME: Steel platform.
PERFORMANCE: Typically 149 mph (240 km/h).
CURRENT PRICE GUIDELINE: ***** One of the most exotic cars.
PROBLEMS TO LOOK FOR: Misuse, old age.
CLASSIC CAR STATUS: ***** One of the greatest.

△ **The Pegaso Z102B became one of the most exotic post-war cars. Owner: Paul Kunkel.**

In the early 1940s, the most advanced machines seen on Spanish roads were huge six-wheeled pre-civil war trucks made by firms such as Leyland in Britain. These monsters, the design of which had hardly changed since the First World War, were lovingly maintained by back-street craftsmen who could make almost any spare part. This was an ideal situation on a limited scale, but hardly conducive to expansion in an impoverished economy. What Spain needed if it was to develop as a nation, rather than as somebody else's satellite, was a national truck – or so thought General Franco.

Consequently, the Instituto Nacional de Industria (Ministry of Industry) created ENASA – the Empresa Nacional de Autocamiones – for that purpose in 1945, hiring a Spanish former Porsche and Alfa Romeo engineer, Wilfredo Ricart, as chief designer. Ricart was well suited to the task because his experience ranged from commercial vehicles to Grand Prix cars and he came from Barcelona, where ENASA had just taken over the old Hispano-Suiza works (see page 77). The name of the truck was to be Pegaso, the flying horse, to show that Spain was really moving into a modern age of transport.

Ricart's work with trucks and buses was so successful that he managed to get Franco's backing for a project which really did appeal to him: building a car better than anything produced by Enzo Ferrari, with whom he had clashed bitterly at Alfa Romeo. Franco saw the project as a way of opening new markets for Spanish technology, and of motivating the workforce at the same time. With such ideals as these, the cost in labour and materials was unimportant.

Ricart, however, was looking for more than a series of dream cars when he started work on the project in 1950. He saw himself as the great white hope of the Spanish car industry. Initially, he talked of producing 200 of these exciting new cars each year, and maybe thousands of simpler ones. So, to power his exotic model, he designed a V8 engine that was in principle very much like his V16 Alfa Romeo Grand Prix design, but which could be split into two four-cylinder units when necessary. For maximum efficiency, it had two overhead camshafts on each bank of cylinders, with chain drive and twin-coil ignition for the road, or gear drive and a magneto . . . just in case Franco wanted a Grand Prix engine.

The influence of Ferrari's new designer, Gioacchino Colombo (see page 65), was not lost on Ricart either, as the engine was 'oversquare', with its bore larger than its stroke to provide maximum revs for efficiency. It also had hemispherical combustion chambers like the brilliant new Jaguar XK (see pages 84-5).

Everything else about the specification of this car – the Z102 – was equally exotic. The dry-sump engine was cast in aluminium, the connecting rods were machined from solid bars, as many as four Weber carburettors were used, and the five-speed gearbox was incorporated in the rear axle in a transaxle like that of one of Ricart's other Grand Prix projects for Alfa Romeo.

The gearbox was also unusual in that it used motor cycle-style dog clutches to allow instant changes without wasting time in dipping the clutch pedal. (This was used merely in starting off.) Everything was beautifully made: the propellor shaft, for instance, was bored on a machine which had been used for making Hispano's famous guns!

The suspension was equally sophisticated for its day – de Dion at the back for the best traction and independent at the front for the best handling. As for the chassis, it was the closest that anybody could come at the time to a supremely rigid monocoque, without compromising the efforts of the world's best coachbuilders who were to provide the bodies.

The first Z102 was shown at Hispano-Suiza's traditional stamping ground, the Paris Salon, in 1951. After that, similar models appeared in a variety of capacities, of which the most popular was the 2.8-litre Z102B of 1953, invariably carrying the finest grand touring or open spider bodywork that frequently put Ferrari to shame. Pegaso even built two with stunning catamaran coachwork. However, as with Lamborghinis a decade later, Pegasos were rarely seen in competition. Franco knew that their stunning appearance, performance and specification was enough to show what Spain could do without their being exposed to possible defeat.

By 1958, when nearly 100 had been built, it was deemed that producing any more would destroy their exclusivity, so Pegaso just plodded along with its trucks and buses, while Ricart went on to become the elder statesman of the Spanish car industry.

SUMMARY

Plus points
Rarer than a Ferrari, more exotic design.

Fantastic craftsmanship.

Excellent performance.

Minus points
Very high restoration costs.

PORSCHE

All Porsche cars have been essentially the creations of either Professor Ferdinand Porsche or of his son, who is also called Ferdinand, but bears the title Dr Porsche. The company has always paid great attention to its development work and now enthusiasts have the widest range of choice available from any sports car manufacturer – open, closed or Targa-topped models, front or rear-engined, normally aspirated or turbocharged, all to basic, medium-priced, or very luxurious specifications.

Some of the earliest machines for which Professor Porsche was responsible included the magnificent supercharged Mercedes-Benz sports cars of the 1920s before he set up his own design agency.

Prof. Porsche, an Austrian, was imprisoned by the French after the war for having helped to design German tanks, and was only freed when his son, known as Ferry, raised a massive ransom by assisting in the design of the Cisitalia Grand Prix car (see page 49). The Porsche family was particularly impressed by the way Piero Dusio based his Cisitalia sports cars around ordinary Fiat components, and decided that it was worth while trying the same idea with the only cars available in any quantity in Austria: old Volkswagens designed by Porsche.

356

YEARS MADE: 1944–55.
ENGINE: Horizontally opposed four-cylinder, overhead valves, pushrods, typically 1086 cc, 40 bhp at 4200 rpm.
FRAME: Steel platform.
PERFORMANCE: 85 mph (137 km/h).
CURRENT PRICE GUIDELINE: * * * Very rare, good examples difficult to locate.
PROBLEMS TO LOOK FOR: Corrosion, non-originality, neglect.
CLASSIC CAR STATUS: * * * Status earned because it is the earliest Porsche.

The first car to bear the name Porsche was, like the Auto-Union and Cisitalia Grand Prix cars, of mid-engined design – that is, with the engine behind the driver but in front of the rear axle. This configuration concentrated the main mass of metal in the centre of the car to very good effect: it gave the best balance, good traction from having a fair degree of weight over the driven wheels, and good aerodynamics because it was possible to have a nicely streamlined nose that did not have to stick up to accommodate a front-located engine. It also eliminated the propellor shaft which would normally connect the engine and gearbox to the rear axle, thereby both saving weight and making the car more responsive,

▽ The lines of the early Volkswagen-derived Porsche 356 were continued for four decades. Owner: Ivor Coverley.

△ ***The classical Porsche Speedster stayed in production for only three years. Owner: Ray Wright.***

since there was not so much slack to take up in the driveline.

The basic Volkswagen components were especially suitable for this layout because the small four-cylinder air-cooled engine was of horizontally opposed (or 'boxer') formation. This concentrated the main weight of metal low down, promoting better handling and allowing far better rearward vision than with a normal engine of vertical format. The only real problem was that, in a mid-engined formation, the engine occupied the space which would normally be given over to rear seats.

This car became the forerunner of generations of mid-engined Porsche sports racing and single-seater competition cars, just as a second prototype became the ancestor of all the company's road cars for the next 27 years.

In the second car the position of the engine reverted to that of the Volkswagen – behind the rear axle line. This made it a rear-engined, rather than a mid-engined car and left enough room at the back for two small passenger seats, or a substantial luggage platform. Luggage could be carried in the nose, but the amount was restricted because the front of the car was low and streamlined. The main problem was that, since so much weight was concentrated behind the rear wheels, the tail tended to swing round like a pendulum once the tyres lost adhesion: this process was made all the more likely by the layout of the typically German swing axle rear suspension. Be that as it may, having so much weight at the back had one notable advantage: it endowed the back wheels with far more traction than normal, which was a definite plus point on the rough mountain roads of the Porsche family's native Austria.

The Porsches saw a good market for sports cars in Europe because, after the war, most manufacturers were concentrating on trying to produce the more popular saloon cars. They therefore decided to build more of the rear-engined cars as these would be more appealing for touring and could be made from basic Volkswagen components with little modification.

One of the problems with the Volkswagen engine was that it was small and therefore did not provide a lot of power, so it was decided to give the new touring car a very aerodynamic body – rather like that of the open mid-engined car and bearing a distinct resemblance to the Volkswagen saloon – to ensure as good a performance as possible. This beautiful body, designed by Erwin Kommenda, was hand-made like the rest of the car: these first Porsches, designated 356 because the model was the Porsche design agency's 356th project, were very expensive.

However, they performed so well that soon a profitable business was established. Many of the early customers were competition drivers who were desperate for any new sports car and who, since they were so skilled, could exploit the tail-happy handling to great advantage.

Soon the Porsche business was going so well – despite the death of Professor Porsche in 1951 – that the team was able to move back to its pre-war headquarters in Stuttgart, West Germany, and modify the basic 356 model to improve its performance. At the same time, Porsche benefited from the increasing American demand for imported cars that had been fostered by the M.G. TC (see page 127). Indeed, the small M.G.s were in such demand that in the early days importers were making a fortune from forcing their agents to take two Volkswagens with every M.G. These Volkswagens were then sold at a discount and the American customers, at first reluctant to buy German products, discovered that the

Volkswagens were extremely well made and reliable, so they bought more – and started taking Porsches, too. Soon America, and particularly fashion-conscious states such as California, became Porsche's biggest market outside West Germany.

SUMMARY

Plus points

Delightful appearance.

Very durable mechanical parts once in good condition.

Excellent steering.

Economical to run.

Numerous events in which it can compete.

Minus points

Prone to corrosion outside California.

Likely to have a hard life.

Expensive to restore if bodyshell is corroded.

Tail-happy handling difficult for tyro.

Some parts rare.

356 SPEEDSTER

YEARS MADE: 1954–7.
ENGINE: Horizontally opposed four-cylinder, overhead valves, pushrods, typically 1488 cc, 70 bhp at 5000 rpm.
FRAME: Steel platform.
PERFORMANCE: 95 mph (153 km/h).
CURRENT PRICE GUIDELINE: **** Much sought-after early Porsche.
PROBLEMS TO LOOK FOR: Non-originality, corrosion, neglect.
CLASSIC CAR STATUS: *** Set to appreciate.

As the mid-engined Porsche competition cars became increasingly successful, the rear-engined sports cars – in open, fixed-head and more luxurious drop-head coupé form – became more and more specialized, using engines of 1100 cc, then 1300 cc, and eventually 1500 cc.

The most potent examples, powered by 1500 cc racing engines, were called Carreras after successes in the Carrera Panamericana road race. During those early years, one of the most successful American dealers, the Austrian-born Max Hoffman, whipped up a demand for a spartan roadster by chopping the top off a competition coupé to reduce weight and provide better visibility on tight race circuits.

This body style was produced chiefly for his home market, and its initial name – the American Roadster – was changed in 1954 to the Speedster. This stark little machine became one of the world's best-loved cars and still symbolizes for many thousands of people, especially Americans, the essence of Porsche motoring. However, it stayed in production for only three years because it did not encourage sales of the optional luxury equipment which has always been one of Porsche's most profitable lines.

SUMMARY

Plus points

Charming appearance.

Very durable mechanical parts once in good condition.

Excellent steering.

Reasonable performance, very good with competition engine.

Economical to run.

Excellent availability of many parts.

Numerous events in which it can compete.

Minus points

Prone to corrosion outside California.

Likely to have had a hard life.

Expensive to restore if bodyshell is corroded.

Tail-happy handling difficult for tyro.

CARRERA RS

YEARS MADE: 1972–3.
ENGINE: Horizontally opposed six-cylinder, 2687 cc, single overhead camshaft, 210 bhp at 6300 rpm.
FRAME: Steel platform.
PERFORMANCE: 150 mph (241 km/h).
CURRENT PRICE GUIDELINE: **** Set to continue recent rapid appreciation.
PROBLEMS TO LOOK FOR: Non-originality, numerous fake examples, corrosion, mistreatment, ultimate European engine illegal in United States.
CLASSIC CAR STATUS: **** Ultimate Porsche 911.

Eventually it was realized that the two-door 356, in A, B and C forms, with engines up to 1600 cc and a screaming 2 litres on the fastest Carrera coupés, was rather too small for modern tastes. This applied particularly to Americans. Porsches had also progressed far beyond using Volkswagen components, although the 356 was of the same basic rear-engined layout. Therefore, work began in the late 1950s to design a larger model: the 901.

Ferry Porsche would not have it made larger than normal grand touring cars, however, in keeping with the time-honoured German cartel system where each manufacturer restricts himself to just one product. Porsche's

▽ Porsche's greatest 911, the Carrera RS, succeeded beyond the maker's wildest dreams. Owner: John Locke.

judgment on the member of his design team who suggested a four-door Porsche was expressed in that now famous quote: 'Shoemaker, stick to your last.'

Consequently, when the 911 (as it had to be renamed after Peugeot claimed that an 0 in the middle of a type name was its trademark) appeared in 1964, it was along the same basic lines as the 356. There were two small seats in the back and an engine – now with six cylinders, based on a Grand Prix design – still hanging behind the rear axle. Happily this conventional assortment was contained in one of the best-looking bodies of all time, designed by Ferry Porsche's son, Butzi.

The 911 was then developed along the same lines as the 356: progressively more powerful and flexible engines were fitted to make new models more attractive than the previous ones. Alternative body styles were also developed, notably an open Targa in which a virtue was made of a built-in roll cage by plating it in stainless steel. In general, the early 911s were sold in three specifications: the cheap and basic Touring on carburettors; the medium-priced and more powerful E (for *Einspritz* – fuel injection); and the high-performance S (for sporting).

During the early 1970s, however, Porsche competition activities with the mid-engined cars became increasingly expensive, culminating in turbocharged models which were soon so dominant in their classes that the rules were changed to eliminate them and give the other manufacturers a chance.

Porsche began to concentrate on production racing cars in the early 1970s, the first of which became the greatest classic: the Carrera RS (for *Renn Sport* – road racing). This car, produced only in 1972 and '73, was a development of the earlier 911S and proved outstandingly popular both in competition and on the road.

SUMMARY

Plus points

Superb appearance, performance, brakes and steering.

Very durable once in good condition.

Excellent availability of parts.

Numerous events in which it can compete.

Minus points

Corrosion.

Tail-happy handling can be difficult for tyro.

Expensive parts and restoration.

930 3.3-LITRE

YEARS MADE: 1978 to date.
ENGINE: Horizontally opposed six-cylinder, 3299 cc, single overhead camshaft, 300 bhp at 5500 rpm.
FRAME: Steel platform.
PERFORMANCE: 161 mph (259 km/h).
CURRENT PRICE GUIDELINE: * * to * * * * According to age.
PROBLEMS TO LOOK FOR: Mistreatment.
CLASSIC CAR STATUS: * * * Set to appreciate.

In 1974, American environmental laws forced the use of far heavier bumpers for all markets (since it was not economically viable to produce separate bodyshells for different countries) and lower-powered anti-emission engines for the United States. As a result the Porsche 911 engine grew and grew to maintain the performance which was constantly being lost due to extra weight and power-sapping regulations. This made the car more expensive, so it was given more luxury equipment to justify its price, which made it heavier still, and so on. . . .

By 1975 the new top model, called the 930 because it was so different from the normally aspirated 911, had a turbocharger that had been under development since 1969. Engine capacity started out at 3 litres and had risen to 3.3 by 1978.

SUMMARY

Plus points

Vivid performance.

Superb brakes and steering.

Very durable.

Corrosion-resistant bodyshell.

Excellent availability of spares.

Minus points

Handling can catch out the unwary.

Top engine not available in United States.

◁ *Porsche's awe-inspiring turbocharged 3.3-litre 911-shaped 930.*

▷ *The front-engined water-cooled Porsche line was extended for the svelte grand-touring 928S.*

928S

YEARS MADE: 1981–6.
ENGINE: V8, 4664 cc, single overhead camshaft, 300 bhp at 5900 rpm.
FRAME: Integral.
PERFORMANCE: 158 mph (254 km/h).
CURRENT PRICE GUIDELINE: *** to *** According to age.
PROBLEMS TO LOOK FOR: Mistreatment.
CLASSIC CAR STATUS: *** Set to appreciate.

In 1975 a new model was waiting in the wings: the 924. This small four-cylinder fixed-head coupé was made in conjunction with Volkswagen and Audi at the old NSU works to a conventional front-engined format as a replacement for an earlier, unsuccessful, low-priced model, the mid-engined 914.

Porsche then introduced a brand-new grand touring car, the 928, in 1977. It was of a similar, although far more sophisticated, front-engined format to the 924, but with a brand new V8 engine because Porsche was convinced that the 911 design – in essence dating back to the Volkswagen of 1936 – could not carry on much longer. The 928 was a fine car, and handled far more forgivingly than the 911, but to Porsche's surprise, the 911 refused to lie down and die. It sold just as well as ever, because so many legends had grown up around its performance on road and track.

SUMMARY

Plus points
Superb engine, performance, handling, quality.

Minus points
Lacks competition image of 911.

Noisy for a svelte GT car.

944 TURBO

YEARS MADE: 1985 to date.
ENGINE: In-line four-cylinder, 2479 cc, single overhead camshaft, 220 bhp at 5800 rpm.
FRAME: Integral.
PERFORMANCE: 152 mph (245 km/h).
CURRENT PRICE GUIDELINE: ****
PROBLEMS TO LOOK FOR: Harsh ride.
CLASSIC CAR STATUS: ***Excellent quality.

▽ ***Porsche's front-engined 924 became the 944, with 911-style performance when turbocharged.***

The 924, like the 911, found itself upgraded, mainly because it was powered by what was essentially an Audi engine, to keep down the cost. Eventually it was developed into the 944, powered by an in-line four-cylinder Porsche engine: one half of that of the 928! As ever, Porsche has continued its development each year, now offering open, closed or Targa-topped models, front or rear-engined, and normally-aspirated or turbo-charged versions.

SUMMARY

Plus points
Great development of original 924 theme with strong frugal engine.

Excellent performance.

Minus points
Lacks 911's race-bred image.

Poor ergonomics.

Steering not ultra-sensitive.

RELIANT

The British firm Reliant started making sports cars almost by accident after cornering the market in tax-saving three-wheeler cars following the Second World War. In its three-wheeler production Reliant became a pioneer in the use of glass-fibre for bodywork from 1956.

SCIMITAR GTE

YEARS MADE: 1968–75.
ENGINE: V6 Ford, 2994 cc, overhead valves, pushrods, 138 bhp at 5000 rpm.
FRAME: Tubular.
PERFORMANCE: 118 mph (190 km/h).
CURRENT PRICE GUIDELINE: * * Set to appreciate steadily.
PROBLEMS TO LOOK FOR: Wear and tear.
CLASSIC CAR STATUS: * * Underrated as an innovator.

In the late 1950s Reliant was approached by the Israeli firm Autocars to help in building a cheap new sports car, the Sabra, because of its expertise in handling glass fibre (much cheaper than steel panels). Part of the deal was that Reliant would put the car into production in Britain as the Sabre.

However, when the sports car appeared in 1961 with a steel chassis and glass-fibre body, Reliant's lack of experience in this new field was only too evident – the car was far from attractive and relatively under-powered with its four-cylinder Ford Consul engine. Still, the Sabre began to sell in small quantities when it was restyled and fitted with a six-cylinder Ford Zephyr engine the following year.

At the same time, industrial designer Tom Karen took over the Ogle studio following the death of its founder, David Ogle. One of his first projects was to restyle another sports car of unconventional appearance, the Daimler SP250. Daimler's new owner, Sir William Lyons of Jaguar, was impressed by Karen's show car, but did not put it into production because it would have cost about the same amount as the Jaguar E-type (see page 89).

The new coupé went begging until Ogle asked Reliant if it would take over production of a rather expensive special-bodied Mini that Ogle was making at the time. Reliant was not interested, but asked whether the Ogle show car's body could be adapted to take the running gear from the Sabre.

The Scimitar which resulted was very much like the Ogle Daimler except that it had a bigger 'greenhouse' – or cabin – to give more headroom, and had been revised to accept Reliant's Ford Zephyr-based mechanical package.

▽ *Reliant's Scimitar GTE started new trends in the 1960s and returned to production in 1987.*

As the Scimitar two-seater coupé was being developed, Ogle revamped Reliant's Robin three-wheeler – which had always sold better than its motor cycle-based rivals because it had a van body – giving it a significant new rear window. In addition, as part of Reliant's new-found speciality of developing new cars for emergent countries, Ogle designed the body for the Anadol, the first car to be built in Turkey.

Ogle also produced an adventurous new show car, the GTS, to take its bow in 1965 alongside the new Scimitar production car. This had a Scimitar body with a new greenhouse which featured bonded glass as a showpiece for the latest technology.

At the time that Ogle was designing his GTS show car, Reliant was working on an amazing project – a high-performance version of the utterly conventional Anadol, for which Ogle had designed a new two-door body. Suddenly, when the GTS and the Robin rear window were looked at in conjunction with the high-performance Anadol body, Ogle and Reliant realized that a new type of car had been invented – a grand touring estate!

Reliant managing director Ray Wiggin liked it immediately, and by early 1968 Ogle had designed and built the prototype with modified Scimitar running gear. This involved widening and lengthening the chassis to liberate extra room for back seats, and setting the latest V6 Zephyr engine well back for good weight distribution. Further innovations included the first rear seat in any estate car which could be folded down in two sections to allow three people to be carried with long items such as fishing rods.

Once the controversial lines of the GTE, launched in October 1968, had been accepted it sold well. One of the teething problems was that the rear window quickly became covered in dirt from the road. This was solved by the fitment of a rear wiper from October 1969: the GTE was the first car to have such a device.

Sales continued to soar, helping to make Reliant the second-largest all-British car manufacturer in 1969. Ogle went on to design the revolutionary Raleigh Chopper bicycle, but was called in again in 1976 to restyle the GTE with a longer wheelbase for more comfort, and to help with a mark two version in 1980.

Soon afterwards, Reliant decided to build a cheap glass-fibre sports car in far greater quantities to fill the void left by the demise of the MGB (see page 129). In this instance the company adopted a design by the Italian firm, Michelotti, but the unconventionally styled Scimitar SS1, as it was called, was nothing like as successful as the classical GTE it replaced in 1984.

SUMMARY

Plus points

Pioneering design.

Comfort.

Corrosion-free body.

Good performance.

Lightweight economy.

Good availability of low-priced spares.

Minus points

Electrical problems inherent with glass-fibre construction.

RENAULT

As the Lancia Stratos took international rallying into the supercar class (see page 104), Renault's tiny R5 Turbo heralded a new era of ever-increasing power outputs. Its conception was the last act of Renault's competitions director, Jean Terramosi, before he retired in 1975.

R5 TURBO

YEARS MADE: 1979–80.
ENGINE: In-line four-cylinder, 1397 cc, overhead valves, pushrods, turbocharged, 250 bhp at 6000 rpm.
FRAME: Tubular and sheet steel.
PERFORMANCE: 130 mph (209 km/h).
CURRENT PRICE GUIDELINE: *** Set to appreciate.
PROBLEMS TO LOOK FOR: Mistreatment.
CLASSIC CAR STATUS: **** Historic homologation special.

When he was planning the R5 Turbo, competitions director Terramosi visualized a ferocious mid-engined sports car with a body resembling the silhouette of Renault's top-selling 5 mini-saloon.

The ideal engine would have been the 2.7-litre V6 from Renault's top saloon, the R30. It offered plenty of power and a great deal of torque which would have made such a small, light vehicle very fast, and very easy to drive. However, Renault was already committed to extending the use of turbochargers as a way of getting something for nothing in the power output race. Quite simply, there seemed to be more future in using an engine's exhaust gases to work a turbine, which would force in more fuel to raise the power output, than in using the old method of increasing the capacity of the engine, which would then weigh more and take up more room.

With these basic thoughts in mind, Renault went ahead with an ambitious programme of sports car and then Grand Prix racing, to develop and publicize turbochargers. The R5 Turbo – with the word 'Turbo' emblazoned along its flanks in bold lettering – was to be the first of a new breed of rally cars that would, hopefully, make customers clamour for mass-produced saloons bearing that magic word.

Although the new R5 Turbo had to look vaguely like an ordinary R5, it would need far larger wheels to convey a great deal more power, so the Italian firm Bertone was entrusted with the task of somehow incorporating huge wheelarches within the Frenchmen's favourite mini-car shape. Inevitably, development was slow, because every time something was changed for technical reasons, the shape had to change as well since there was so little room inside. In principle, however, there was no hesitation in the designers' minds. The car had to be the most proficient road racer imaginable, with proper racing car suspension – a double wishbone independent system at each corner. As the engine would have to be turbocharged, the existing 1397 cc Renault Alpine power unit was chosen.

This sporting model became the first French petrol-engined car to be fitted with a turbocharger: an exhaust-gas-driven version of the mechanically driven supercharger which was so popular before the war, having been developed on diesel-engined trucks. Superchargers were considered, but their mechanical drives absorbed too much power on a small, light engine.

▽ The extraordinary Alpine-built Renault R5 Maxi Turbo, in its element at the Rally d'Antibes.

Both four-wheel-drive and front engine/rear transaxle designs were considered before a mid-engined location was selected on the grounds that it gave the best weight distribution.

No other calculable factor was left to chance: even the petrol tank was situated as low as possible, under the seats of the car's two occupants, so that the changes in weight, as it was emptied or filled, would have the minimum influence on roadholding.

The brakes featured huge ventilated discs on each wheel to combat overheating, and the wheels themselves were made in differing sizes, front and rear, to carry the latest high-speed Michelin TRX tyres. The resulting grip on the road was phenomenal.

As this was a serious competition-oriented exercise, there was originally little thought of selling the car to the general public. The R5 Turbo first appeared as a prototype at the Paris Motor Show in 1978 but, with its lightweight body, compact dimensions and advanced specification, this potential projectile set enthusiasts' hearts racing. Their dreams were realized in 1979 when the 400 examples needed to homologate – or qualify – the R5 Turbo for class four international rallying, started to leave Renault's recently acquired Alpine factory (see page 19).

Soon after the requisite number had been built, Jean Ragnotti celebrated by taking an R5 Turbo to victory in the 1981 Monte Carlo Rally. The day of the turbocharged supercar had dawned; but the R5 Turbo's reign at the top was short. Close on its heels came Audi's new Quattro (see page 25), which – with four-wheel drive – went on to dominate rallying in the years to follow.

SUMMARY

Plus points

- Fantastic handling.
- Enormous grip.
- Very exciting performance.
- Excellent brakes.

Minus points

- Very noisy.
- Extremely cramped accommodation.
- No luggage space.
- Extremely spartan décor.

ROLLS-ROYCE

Rolls-Royce, whose cars have been awarded the accolade of 'The Best Car in the World', has always been a typically British institution, derived initially from a partnership between electrical engineer Henry Royce and car salesman Charles Rolls.

Henry Royce's first car was a noisy and unreliable 1901 Decauville and, being a perfectionist, he was sure he could do better.

Using the Decauville's design as a basis, he built his own version – and two others – in 1904. It was immediately obvious that these were excellent machines, and their reputation soon reached the ears of the Hon. C.S. Rolls.

The patriotic Rolls, who until then had been forced to trade mostly in imported machinery, was highly impressed and offered to sell everything Royce could make. Three of the new cars, bearing the name Rolls-Royce, were then exhibited on Rolls's stand at the Paris Motor Show in December 1904.

Rolls, a pioneer competition driver, enjoyed a great deal of success with these early cars, including winning the Tourist Trophy race in 1906. He also provided the basis for publicity stunts by advertising man Claude Johnson, who became known as 'the hyphen in Rolls-Royce'. These stunts included taking the world reliability record with 7089 miles (11,406 km) non-stop in 1907 in the first really great Rolls-Royce car, the six-cylinder side-valve engined 40/50, later to be called the Silver Ghost.

SILVER GHOST

YEARS MADE: 1906–26.
ENGINE: In-line six-cylinder, side-valve, typically 7428 cc, power not revealed.
FRAME: Twin girders, cross-braced.
PERFORMANCE: Typically 60 mph (97 km/h).
CURRENT PRICE GUIDELINE: ***** Gilt-edged investment.
PROBLEMS TO LOOK FOR: Non-originality, especially of body.
CLASSIC CAR STATUS: ***** One of the greatest.

△ ***The splendid Rolls-Royce Silver Ghost was christened because it ran so silently. Owners: D. K. and R. A. Lankester.***

The 40/50 first shown in 1906 – set new standards in silence of operation, seemingly unlimited power, supreme reliability, superb steering, and excellent brakes (by the standards of the day). It put the chassis – Rolls-Royce did not make its own bodywork until after the Second World War – into a class of its own and promptly received the title of 'The Best Car in the World'.

One of Johnson's most successful stunts was to have the 40/50 Motor Show chassis clad in silver-painted and plated bodywork and christened the Silver Ghost, because it was so silent.

Demand for the Silver Ghost was so high that Rolls-Royce had to move from Royce's old factory in Manchester to a new works at Derby in 1908. Rolls, in the mean time, extended his sporting activities to flying, becoming the first man to fly to France and back non-stop. He was the first Englishman to be killed in a flying accident.

Johnson then took over much of the running of the business as Royce's

SUMMARY

Plus points

Magnificent engineering.

Superb reliability.

Surprisingly good performance.

Usually splendid coachwork.

Great comfort in good weather.

Minus points

High cost of restoration and spare parts.

PHANTOM I SPRINGFIELD

YEARS MADE: 1926–31.
ENGINE: In-line six-cylinder, 7668 cc, overhead valves, pushrods.
FRAME: Twin girders, cross-braced.
PERFORMANCE: Typically 85 mph (137 km/h).
CURRENT PRICE GUIDELINE: **** Relatively undervalued.
PROBLEMS TO LOOK FOR: Non-originality, especially of body.
CLASSIC CAR STATUS: *** Set to appreciate.

health deteriorated after years of overwork. From 1911, much of Royce's design work would be carried out from his villa in the South of France.

Thus the 40/50 continued to be modified, but with no major changes, until 1925. Production continued on a reduced scale throughout the First World War, when many 40/50s made magnificent armoured cars weighing more than 6 tons (6000 kg). Numerous other makes were used but, by general agreement, there was nothing to match a Rolls-Royce for armoured cars.

Under pressure from the government, during the First World War Rolls-Royce began to build 12-cylinder aero-engines. Like the Silver Ghost, these proved so reliable that they had a great effect on the British war effort, later powering the first 'plane to cross the Atlantic. However, after the war, the demand in Britain for very expensive luxury cars and aero-engines was low and Johnson decided to open a subsidiary factory in Springfield, Massachusettes, to cater for the American market. The Silver Ghost was made there until 1926, usually complete with a body built by Rolls subsidiary Brewster of New York, as not even Johnson could persuade the Americans that it was normal to purchase a car chassis, then take it to a coachbuilder to have a body fitted. Strangely enough, they preferred to buy a complete car from the showroom!

To counter the drop in home sales of the 40/50, which continued in reduced production at Derby, Rolls-Royce introduced a new, smaller, model called the Twenty, which became very popular from 1922. In 1924 this became the first Rolls to be fitted with four-wheel brakes.

The new overhead-valve engine which had been designed for the

▷ ***The American market forced Rolls-Royce to produce its own Springfield coachwork. Owner: William Northrup.***

△ ***Bentley's 8-litre was such a threat to Rolls-Royce's Phantom II that the rival firm was taken over. Owner: Tom Mason.***

Twenty formed the basis in 1925 for the power unit of a revised version of the 40/50, called the New Phantom later called the Phantom I). Production at Springfield was switched to the Phantom I in 1926. Johnson died in the same year, but Royce soldiered on with a good management team, replacing both the Phantom I and the Twenty in 1929. Construction of Rolls-Royce cars at Springfield ceased in 1931.

SUMMARY

Plus points

Magnificent engineering.

Superb reliability.

Extremely good performance.

Great comfort in good weather.

Typically more inspiring coachwork than on British Phantom Is.

Spare parts position better than Ghost.

Minus points

High costs of restoration.

PHANTOM II

YEARS MADE: 1929–35.
ENGINE: In-line six-cylinder, 7668 cc, overhead valves, pushrods, power not revealed.
FRAME: Twin girders, cross-braced.
PERFORMANCE: Typically 92 mph (148 km/h).
CURRENT PRICE GUIDELINE: **** Set to appreciate.
PROBLEMS TO LOOK FOR: Non-originality, especially of body.
CLASSIC CAR STATUS: **** Set to appreciate.

The Phantom's chassis, the ancestry of which could be traced back to that first Silver Ghost in 1906, was lowered, with half-elliptic springing all round and renamed the Phantom II. The Twenty had its engine enlarged, and with chassis modifications, became the 20/25.

Royce's design team at the time included Ivan Evernden, who was to help create the Bentley Continental after being responsible for the most attractive 'Continental' bodies built by Barker on the Phantom II chassis. Rolls-Royce was able, in 1931, to buy the bankrupt Bentley Motors, and stop rising competition from its 8-litre.

SUMMARY

Plus points

Magnificent engineering.

Superb reliability.

Extraordinary performance.

Better engine than Phantom I.

Great comfort.

Spare parts position better than earlier models.

Minus points

Typically uninspiring coachwork if original.

High costs of restoration.

PHANTOM III

YEARS MADE: 1936–9.
ENGINE: V12, 7340 cc, overhead valves, pushrods, power not revealed.
FRAME: Twin girders, cross-braced.
PERFORMANCE: Typically 92 mph (148 km/h).
CURRENT PRICE GUIDELINE: ***** One of the all-time greats.
PROBLEMS TO LOOK FOR: Non-originality, engine wear in early models.
CLASSIC CAR STATUS: ***** Gilt-edged investment.

Rolls-Royce's ever-increasing involvement with multi-cylindered engines for aircraft and ships was reflected in the Phantom III of 1936. Its V12 power unit owed much to aero-engine design and developed approximately 170 bhp at 3000 rpm: as ever, Rolls-Royce refused to reveal its car's power output, merely describing it as 'adequate'. There were considerable changes to the chassis, too, involving the use of the General Motors' system of independent front suspension. This had been developed largely by Maurice Olley, formerly of Rolls-Royce's now-obsolete Springfield factory!

SUMMARY

Plus points

Magnificent engineering.

Superbly smooth engine.

Amazing performance.

Splendid coachwork.

Great comfort.

Minus points

Engine wear in some early models.

Very high costs of restoration.

SILVER CLOUD III

YEARS MADE: 1962–6.
ENGINE: V8, 6230 cc, overhead valves, pushrods, power not revealed.
FRAME: Twin girders, cross-braced.
PERFORMANCE: 115 mph (185 km/h).
CURRENT PRICE GUIDELINE: *** Set to appreciate.
PROBLEMS TO LOOK FOR: Neglect.
CLASSIC CAR STATUS: ** Set to appreciate.

One of the most favoured Rolls-Royce coachbuilders had been Barker, which specialized in heavy formal limousines, but when Park Ward, which normally produced lighter, more sporting, bodies, ran into financial trouble in 1938, it was taken over to act as the company bodybuilder. As the demand for aero-engines increased in the years immediately before the Second World War, Rolls-Royce was already preparing to switch the emphasis of car

◁ Rolls-Royce used aero-engine technology in the Phantom III. Briggs Cunningham Museum.

▽ The Silver Cloud Series III with transatlantic headlamps. Merle Norman Classic Beauty Collection.

△ ***The drop-head Rolls-Royce Corniche was so attractive that it sold for a premium.***

production to the less labour-intensive smaller models, using ready-made bodies. It was in this climate that the Wraith, developed from the 25/30, was introduced in 1939.

Rates of pay soared during the war, and it was no longer possible to base production on tremendously labour-intensive cars. Consequently, a standard steel body was adopted, with an overhead-inlet/side-exhaust military adaptation of the earlier six-cylinder engine.

At first, the factory felt embarrassed about calling these cars Rolls-Royces, partly because it had no control over the specification of the steel available immediately after the war. As a result, when production began in 1946, all cars bore the Bentley badge.

There was a strong demand, however, in the more prosperous American market, for 'a proper Rolls-Royce', so the Silver Dawn model was created from a Bentley standard steel saloon basis in 1947. In this case an optional automatic gearbox, based on the General Motors' Hydramatic (see page 138), was fitted. The basic chassis was available, however, for the use of whatever specialist coachbuilders were left, as the Silver Wraith. This chassis then remained unchanged, in essence, until the end of the next series of models, the Silver Cloud and Bentley SI in 1960, taking in the Bentley Continental R type.

The Rolls-Royce standard bodywork was restyled as the Silver Cloud (and Bentley SI) which continued to make up the bulk of production – with the purchase of coachbuilders H. J. Mulliner consolidating body supplies – before a new V8-cylinder engine was introduced in 1959. This was a completely modern unit with overhead valves which has remained in production into the 1980s.

By the time these series II versions of the Silver Cloud and Bentley S-series had been introduced, refinements from America, such as power-assisted steering, had become standard fittings. Twin headlamps were also 'imported' for the series III and with the same running gear, from 1962.

SUMMARY

Plus points

- Magnificent engineering.
- Very attractive bodywork.
- Great comfort.
- Superbly smooth engine.
- Excellent reliability.
- Excellent availability of spare parts.

Minus points

- Corrosion in early models.
- Handling less refined than a Jaguar.
- High costs of restoration.

CORNICHE

YEARS MADE: 1971 to date.
ENGINE: V8, 6750 cc, overhead valves, pushrods, power not revealed.
FRAME: Integral.
PERFORMANCE: 120 mph (193 km/h).
CURRENT PRICE GUIDELINE: *** to **** According to age.
PROBLEMS TO LOOK FOR: Neglect in early models.
CLASSIC CAR STATUS: **** One of the greats.

By the mid-1950s, the motor industry generally had switched to integral construction, in which the chassis is combined with the bodyshell for more efficient construction, lower lines and less wind resistance. Rolls-Royce could not put off the move any longer and, following a long period of development, a completely new type of Rolls-Royce appeared (Bentley-trimmed production now being around 5 per cent of the total). This new model was the Silver Shadow of 1966, which was to become the best-selling Rolls-Royce.

The well-tried V8 engine was retained, with new independent rear suspension which improved the ride and handling, in a shorter and lower bodyshell of far more modern design.

In 1971, the special-order Silver Shadows were developed into the Corniche, which had a lower roof line than the Shadow and a deeper radiator shell.

Rolls-Royce ran into severe technical difficulties (and, as a result, financial problems) with advanced new materials in the RB211 aero-engine, and these troubles led to nationalization. The profitable motor car, diesel and light piston engine division was then sold off to go public as a separate company in 1973 and has continued to prosper since, with a very expensive coachbuilt Pininfarina-styled Carmargue offered from 1975. Rolls-Royce was taken over by Vickers in 1980 and introduced the lower and wider Silver Spur as the standard model in 1981. In that year the Bentley model name was revived and now, with the very fast Mulsanne Turbo (see page 34) and the cost-cutting 8, it accounts for a far more substantial proportion of total output.

SUMMARY

Plus points
- Magnificent engineering.
- Superbly smooth engine.
- Great reliability.
- Decent performance.
- Great comfort.
- Excellent availability of spare parts.

Minus points
- Corrosion in standard steel bodywork.
- Rather clumsy handling.
- High costs of restoration.

STUTZ

The greatest sports cars ever made in America bore the name of Harry C. Stutz, an Ohio farm boy of Dutch descent who was the original 'shade-tree mechanic' . . . he could fix anything!

Stutz relied more on a dogged determination to do everything just right, than on any formal education. Although he often fiddled with cars, it was not until 1911 that he actually built a complete one, and then only as a test bed to demonstrate a rear axle he was trying to sell. However, like everything else Stutz built, it was so well made that when he entered it at the local track – Indianapolis – it finished 11th against seasoned racers.

BLACK HAWK SPEEDSTER

YEARS MADE: 1927–35.
ENGINE: In-line eight-cylinder, typically 5277 cc, single overhead camshaft, supercharger, 143 bhp at 3800 rpm.
FRAME: Twin girders, cross-braced.
PERFORMANCE: 105 mph (169 km/h).
CURRENT PRICE GUIDELINE: ***** One of America's greatest.
PROBLEMS TO LOOK FOR: Non-originality, simple old age.
CLASSIC CAR STATUS: ***** All-time classic.

The first Stutz was powered by a Wisconsin four-cylinder engine that was not so refined as the units in the rival Mercers or Marmons but, like the rest of the car, it was rugged and dependable. Stutz also kept an iron grip on his drivers, insisting that they used only standard cars and drove for a finish, rather than that they risk retiring in a mad burst of glory with a fragile, highly tuned car. As a result, Stutz cars gained an enviable reputation for reliability.

Bearcat and Bulldog were among the model names used during this early period for what were pure sports cars which could hold their own in any contest. A record-breaking coast-to-coast run by a former motor cyclist, called Cannonball Baker, aroused so much interest that the factory was swamped with orders and Stutz decided to refinance the operation so that he could step up production from 500 cars a year to 3500.

It was then that Harry Stutz ran into trouble. When he offered his business stock on Wall Street, it was bought up by a financier, Alan A. Ryan, who forced up the price by methods which caused a national scandal and resulted in dealings in Stutz shares being suspended on the New York market.

▽ ***The Stutz Black Hawk Speedster, one of the greatest American sports cars, finished second at Le Mans in 1927.***

Stutz had to quit in 1920 and went off to build the cheaper, and largely unsuccessful, H.C.S. car before dying in 1930. The firm which bore his name was bought by Bethlehem Steel millionaire Charles M. Schwab and some bankers who knew nothing about making motor cars. They stopped the Stutz racing programme, with the result that sales plunged. In 1925 it became clear that a new model was needed to replace what was essentially pre-war machinery.

People who approached the bankers with new creations included the engineers James Scripps Booth and Frederic Ewan Moskovics, who had independently produced highly refined versions of the American Underslung car designed by Stutz in 1905! Moskovics came up with the better proposition and was given the backing to hire a team of engineers, who turned out to be brilliant, to develop his 'Vertical 8 with Safety Chassis' – so called to distinguish it from Packard's straight-eight.

This new Stutz had all sorts of advanced features, including an overhead camshaft, dual ignition, nine main bearings, forced-feed lubrication, underslung worm-drive rear axle, hydraulic brakes and low-line chassis.

As sales soared, so did power outputs, until a climax was reached in the Black Hawk Speedster of 1927. One of these cars managed to lap the Indianapolis Speedway at 71.36 mph (114.84 km/h) for 24 hours without missing a beat . . . and another finished second to a Bentley at Le Mans.

In 1929, Stutz returned to Le Mans with three examples of a 5.2-litre version of the Black Hawk, one of which had a supercharger to give it no less than 143 bhp. Two of the cars had to withdraw, while the remaining one finished fifth.

SUMMARY

Plus points

Fabulous performance for its day.

Excellent handling and brakes.

Magnificent appearance.

Superbly made.

Very durable.

Minus points

Very high restoration costs.

BEARCAT

YEARS MADE: 1930–5.
ENGINE: In-line eight-cylinder, 5277 cc, twin overhead camshafts, 115 bhp at 3900 rpm.
FRAME: Twin girders, cross-braced.
PERFORMANCE: 100 mph (161 km/h).
CURRENT PRICE GUIDELINE: *****
One of the all-time greats.
PROBLEMS TO LOOK FOR: Non-originality, simple old age.
CLASSIC CAR STATUS: *****
Wonderful old-timer.

In 1930, America's greatest sports car ventured back to Le Mans, with what was to be the last of the Stutzes – a new Bearcat model named the DV32 to denote its twin-overhead-camshaft 32-valve cylinder head. Although both entries failed to finish the course, their performance made a lasting impression since one of the cars averaged a staggering 103 mph (166 km/h) for six hours. The eight-cylinder twin-overhead-camshaft DV32 was Stutz's attempt to match the charismatic V12s that had the American market in their thrall.

By now, however, the cost of a Stutz was on a par with that of the Duesenberg it resembled in so many ways, and few people could afford such extravagant machines. In 1935, as sales declined, Stutz, like so many other great marques, went to the wall.

SUMMARY

Plus points

Magnificent performance.

Quality.

Durability.

Minus points

Very high restoration costs.

▽ ***The Stutz Bearcat, with twin-overhead-camshaft V12 engine, a most formidable car.***

△ ***Triumph's TR3A with its 'dollar-grin' grille, one of the most popular British sports cars in America.***

TRIUMPH

Triumph was famous for motor cycles until 1923, when it added cars to the range – mainly good-quality family machines. There was one attempt to make a classic sports car, when a consignment of motor cycles for the Italian army was exchanged for the rights to copy Alfa Romeo's 2.3-litre supercharged sports car in 1935. There was insufficient demand, however, and eventually the car side of the business ran out of money in 1939. The time of the Triumph sports car was yet to come.

TR3A

YEARS MADE: 1957–61.
ENGINE: In line four-cylinder, 1991 cc, overhead valves, pushrods, 100 bhp at 5000 rpm.
FRAME: Twin girders, cross-braced.
PERFORMANCE: 104 mph (167 km/h).
CURRENT PRICE GUIDELINE: ** Set to appreciate.
PROBLEMS TO LOOK FOR: Corrosion, misuse, neglect.
CLASSIC CAR STATUS: ** Set to appreciate.

Triumph was taken over after the war by Sir John Black of the Standard Motor Company who again produced upmarket family saloons, with a limited run of sporting roadsters which were distinguished by having the old-fashioned American-style 'dickey' seat in the tail. Their sales were sufficiently encouraging for Black to commission a far more ambitious touring car, the TRX . . . only to cancel it because it would have cost too much. He then tried to take over Morgan as a short cut to having his own sports car, but failed (see page 133).

So Black asked his design team for a cheap and simple car that could give Standard a share of the burgeoning US market for sports cars that had been created by M.G. and creamed off by Jaguar. The new car would be called a Triumph because the meaning of the name Standard had changed over the years from being the standard by which all others were judged to something relatively inferior in a world in which everything was 'de luxe'. In fact, Black did not really care what the new Triumph looked like, providing that it was cheap, and he offered only the paltry sum of £16,000 for special tools to make the body.

Within the restrictions of this budget, engineer Harry Webster and stylist Walter Belgrove worked wonders to produce an attractive prototype from parts already made by Standard, and cut-price body panels which could not feature any form of double curvature. Consequently, the Triumph 20TS was very cheap at only three-quarters the price of the far more curvaceous Healey 100, launched at the same time in 1952 (see page 28).

Former racing engineer Ken Richardson so completely reworked the 20TS under the skin, making it as fast as the Austin-Healey 100, that his new car was called the Triumph TR2 when it went into production with modified Standard Vanguard 2-litre running gear in 1953. This rugged, bargain-priced sports car was an instant success, particularly in America, and dominated rallies, too.

The face-lifted TR3 version of the TR2 scored a technological first in 1956 when it became the first mass-produced car to use disc brakes. The price stayed amazingly low, however, because hardly anything was spent on developing the body, engine or chassis.

In 1957 the TR3 acquired a full-width radiator grille, and uprated engine as the TR3A. Initially this model was destined solely for the US market, but in early 1958 it became more widely available.

The influential American dealers demanded – and got – a re-run of the wider-grilled TR3A (dubbed the TR3B) after a new body was produced at last on a more expensive TR4 in 1961, to compete with a new breed of more luxurious sports cars, such as the Sunbeam Alpine, which had wind-up windows. The Triumph designers also invented the 'Surrey' top on the TR4, which combined the attributes of an open car and a coupé, but which was too far ahead of its time to be appreciated by many people except Porsche: the German firm copied it five years later and doubled its market with the 911 Targa (see page 144).

In 1962 Triumph rebodied its small Herald saloon – one of the last cars to be made with a separate chassis – as the outstandingly pretty Spitfire sports car. This model provided Austin-Healey and M.G. with stiff competition for their Sprites and Midgets (see page 125) as the firm was taken over by the Leyland truck group, later to become British Leyland.

To improve the comfort on the TR, Triumph at last changed the back suspension to an independent system for the TR4A in 1965 before supplies of the old Vanguard four-cylinder engine ran out – to Morgan's consternation (see page 133). The unit was changed to a modified version of the longer six-cylinder engine from the Standard Vanguard's replacement, a saloon called the Triumph 2000. The TR5 (TR250 in America) which resulted was then rebodied as the TR6 at minimum cost and produced until a completely new car became imperative.

This was the Triumph Dolomite saloon-based four-cylinder TR7 which suffered badly from controversial Lancia Stratos-inspired wedge-shaped styling (see page 104). By the time that the car had received the Rover V8 engine as an open TR8 for the American market, British Leyland had lost so much money on its saloon car operation that economy cuts had to be made and, in 1981, sports car production was shut down.

SUMMARY

Plus points

- Excellent performance.
- Very rugged.
- Superior disc brakes.
- Wonderfully versatile gearbox.
- Easy to restore.
- Low-priced spares.
- Numerous events in which it can compete.

Minus points

- Corrosion.
- Difficult handling when extended in the wet.

TVR

The original TVR cars were the realization of a dream for garage mechanic Trevor Wilkinson: they were his own cars. The marque has gone through some rough patches in its time, and it is a tribute to the imagination, versatility and style of the cars that it has survived against the odds.

TUSCAN V8

YEARS MADE: 1967–70.
ENGINE: V8 Ford, 4727 cc, overhead valves, typically 271 bhp at 6500 rpm.
FRAME: Tubular.
PERFORMANCE: 155 mph (249 km/h).
CURRENT PRICE GUIDELINE: * * * Appreciating.
PROBLEMS TO LOOK FOR: Mistreatment.
CLASSIC CAR STATUS: * * Underrated.

Initially, TVRs were marketed as a multi-tubular chassis into which almost any small engine and transmission could be fitted and then clothed with one of a number of glass-fibre bodies

▷ ***TVR's fearsome Griffith was developed into the far more civilized Tuscan, complete with the same powerful V8 engine.***

which were available commercially. By 1958 these specials had developed into a kit car bearing an outward resemblance to the new Lotus Elite (see page 109). This ultra-short-wheelbase two-seater fixed-head car, called the Grantura, used a Volkswagen Beetle-based system of all-independent suspension and was often fitted with an engine from an MGA (see page 128), or even from the Lotus Elite.

The basic body and chassis style were to remain unchanged for several years as the emergent Blackpool-based company went through a series of bankruptcies and management upheavals brought about mainly by the inability of production to keep up with potential orders, especially American ones. Notable technical changes during the early 1960s included a switch to modern wishbone and coil springs.

The original TVR company's final collapse came in 1964, when a deal with Griffith Inc to install Ford's V8 engine in America for a model called the Griffith V8 fell through. This fearsome machine then became the TVR Griffith, still using the MGB brakes from the far lower-powered Grantura, and a different-bodied version of this car was built around an Austin-Healey 3000 (see page 30) chassis by a firm in Suffolk as the Trident.

Stability returned with the formation of TVR Engineering under new management in 1965. The Grantura was revamped at this stage as the Vixen with a sharply cut-off aerodynamic tail. Most of these cars featured the Ford Cortina GT running gear (as used in the Morgan 4-4, see page 131) or the power train from the MGB (see page 129).

The higher-powered TVR Griffith developed into the more civilized TVR Tuscan, which then received the V6 engine from Ford's Capri that was shared with the Reliant Scimitar GTE. The in-line Triumph engine from the TR5 or TR6 was also used in this chassis, emphasizing its versatility. In 1967 the Tuscan was given the American Ford V8, and the model continued to be produced in this form until 1970, when engine supplies dried up.

Once TVR was on a secure financial footing in the early 1970s, it was able to introduce a new M-type car with a more advanced and longer-wheelbase tubular backbone chassis and modified body. The power trains centred on the Ford V6 or Cortina GT unit, or on Triumph's in-line six-cylinder, which met the demanding US exhaust emission regulations. It was during this period, in 1973, that TVR had to abandon kit car production and concentrate purely on assembled cars because of changes in British taxation laws. This inevitably meant that prices rose, and the model policy had a parallel in Lotus in that it began to be far more up-market. A new top model, the Turbo, was introduced in 1977, featuring a turbocharged version of Ford's V6 engine. This gave it a similar performance to the old Tuscan V8. In the same year, the bodies were restyled in detail, with the addition of a Taimar hatchback and convertible.

The Lotus influence continued to be seen when major changes were revealed in 1980. They centred on a dramatic new wedge-shaped body designed by Oliver Winterbottom, who had earlier been involved in the contemporary Lotus Elite and the Eclat (see page 112), with a fresh tubular chassis by Ian Jones, who came from the same company. This new car, called the Tasmin, and initially powered by the latest Ford Capri 2.8-litre V6 engine, was available in fixed-head, drophead and two-plus-two forms.

More power was then provided by the option of fitting Rover's V8 Vitesse engine from 1984 in the TVR350i – available only in convertible two seater form – with further variants available from 1986 in 3.9-litre and 4.2-litre versions of the V8, called the 390i or the 420SEAC (named after the special composite material used for its body).

The 420SEAC was then developed in the same year into a two-plus-two seater Saloon to compete with Aston Martin's V8 as repeated demands for the traditional rounded-shape TVRs were met by a new low-priced TVR S.

SUMMARY

Plus points

Stunning performance.

Attractive styling.

Relatively durable.

Strong mechanics.

Good availability of spare parts.

Minus points

Wild handling.

Poor original build quality.

VOLKSWAGEN

The name Volkswagen was coined by Hitler in the days of the Third Reich. Literally, it means 'people's car' – a car for the masses; and a car for the masses is what the Volkswagen Beetle turned out to be. It has now been in production for more than 45 years and is recognized instantly by people all over the world.

Although Volkswagen tried to extend its range in later years, it is only relatively recently, with the Polo and the Golf, that the public has welcomed any alternative from Volkswagen to the much-loved Beetle.

△ ***Volkswagen's Beetle reached an ultimate form as the European Cabriolet of 1978. Patrick Collection.***

BEETLE

YEARS MADE: 1941 to date.
ENGINE: Horizontally opposed four-cylinder, overhead valves, pushrods, typically 1285 cc, 44 bhp at 4100 rpm.
FRAME: Steel platform.
PERFORMANCE: 75 mph (120 km/h).
CURRENT PRICE GUIDELINE: * Set to appreciate.
PROBLEMS TO LOOK FOR: Neglect and mistreatment, eventual corrosion.
CLASSIC CAR STATUS: ***** All-time bargain.

The Volkswagen Beetle, of which more than 20 million have been built, making it the most popular car in the world, was essentially the creation of three men: German dictator Adolf Hitler, design agency leader Dr Ferdinand Porsche, and manager Heinz Nordhoff.

The car, which became known as the Beetle in Britain and the Bug in the United States, was in essence Hitler's idea. He set great store by the biography of Henry Ford – read during a term of imprisonment in 1924 – so that when he came to power in 1933 he had two strong objectives: to mop up Germany's unemployed by putting them to work on a massive new road system based on the Italian *autostrade*, and to sell the nation cheap standardized transport in the shape of a *Volkswagen*.

He insisted that it should have a simple air-cooled engine and reflect nature in its body shape – pointing to beetles as a perfect example of efficient streamlining. Hitler's fellow Austrian, Ferdinand Porsche, was instructed to turn the idea into a cheaply produced reality, and this he eventually did when another Austrian, Franz Xavier Reimpress, came up with a four-cylinder horizontally opposed engine for less than the price of the twin-cylinder mechanism which was under development in 1934. He also designed the VW badge which would later become world-famous. Porsche patent torsion bar suspension with a strong steel platform for the body and rear-mounted engine made up the rest of the prototypes.

By 1938, the Volkswagen Beetle looked very much like the examples still being built in Central and Southern America today.

Work began immediately on a massive new factory capable of producing up to 1,500,000 cars a year, and it was just about to begin production when Hitler invaded Poland and triggered the Second World War.

A pilot run of Volkswagen cars was completed in 1941, but almost all the wartime production was of military variants.

When peace came, the British, in whose sector the factory and workers' housing fell, named the settlement Wolfsburg, after a nearby castle. Unfortunately for the British motor industry, the Volkswagen design was turned down by the Rootes group as being out-of-date . . . and the Germans were left to begin producing Volkswagen saloons under great difficulties. Export models eventually found their way to the United States and acquired such a reputation for economy and reliability that massive sales figures were achieved under the guidance of former Opel executive Heinz Nordhoff. By the time that Nordhoff died in 1968, more than 1 million a year were being produced, making Volkswagen one of the world's biggest car manufacturers.

The sheer success of the Beetle led Volkswagen into the peril that nearly trapped Henry Ford when he kept his Model T in production for too long. The Beetle was becoming too dated to enjoy sufficiently large sales, yet any other car that Volkswagen tried paled into insignificance alongside the distinctive Beetle. So it was that the 1500 in 1961, the 411 in 1968 and the front-wheel-drive K70 – which broke new ground for Volkswagen in 1970 in that it was water-cooled – were relative failures. By 1974, the unthinkable was happening: Volkswagen was making a loss!

Relief was at hand, however. In 1965 Volkswagen had acquired Auto-Union, which included NSU and Audi, from Mercedes. After a series of less-than-successful Volkswagen versions of the Auto-Union designs, and the ill-fated mid-engined Porsche 914 venture (see page 145), the company was about to come good with the small hatchback Polo, née Audi 50.

SUMMARY

Plus points

Very durable and reliable.

Amazing traction.

High-quality manufacture.

Timeless design.

Very low running costs.

Ready availability of most spares.

Minus points

Lack of stability in side wind.

Tail-happy handling.

▽ ***Giorgetto Giugiaro's original Volkswagen Golf is still the classic hatchback saloon.***

GOLF GTi

YEARS MADE: 1976 to date.
ENGINE: In-line four-cylinder, single overhead camshaft, typically 1781 cc, 112 bhp at 5500 rpm.
FRAME: Integral.
PERFORMANCE: Typically 119 mph (191 km/h).
CURRENT PRICE GUIDELINE: ** to *** Depending on age.
PROBLEMS TO LOOK FOR: Eventual engine wear and corrosion.
CLASSIC CAR STATUS: *** Set to appreciate.

In 1974, Volkswagen introduced a new front-wheel-drive car styled by Giorgetto Giugiaro. The new car had nothing in common with the Beetle apart from the fact that it had four seats and the same wheelbase. This larger Golf (Rabbit in America) was such a brilliant design that a million were sold within three years of its launch.

The high-performance GTi version produced from 1976 started a new trend for 'hot hatchbacks' which all but wiped out the traditional sports car because the modern bodyshell was lighter and gave it a better performance.

Other models were reasonably successful, but Volkswagen took a really big gamble by restyling the Golf as a more bulbous Series two version in 1983, offering more room in the back. Fortunately the model was so popular by then that it survived without its earlier slim lines. The most recent versions of the Golf include a four-wheel-drive Synchro.

SUMMARY

Plus points

Excellent performance, handling, comfort, economy.

Styling on Series one model.

Minus points

Relatively poor brakes.

Limited rear seat room on Series one model.

More bulbous styling on Series two model.

Eventual corrosion.

ACKNOWLEDGEMENTS

The publishers wish to thank the following for supplying the photographs in the book:

Neil Bruce 22, 80-81; Citroën UK Ltd 53; Hugh Conway Collection 41 above; Mirco Decet 116-117; Chris Harvey 8, 9, 20, 21 above, 25, 28, 29, 30, 76, 95, 98, 99, 100, 104, 110, 111, 113, 130, 132, 133, 134, 147, 152, 156-157; DPPI (Jean Paul Caron) 42; Paul Kunkel Photography 49; The National Motor Museum, Beaulieu: 18, 23, 38-39, 61, 91, 93, 96, 102, 107, 112, 114, 117, 146, 153, 155, 159: (Nicky Wright) 47, 59; Andrew Moreland 58-59, 92, 103, 108, 154-155; Quadrant Picture Library: (Autocar) 82, 90: (Classic Car) 16-17, 115: (Motor) 46; Renault UK Ltd 19; Peter Roberts 55, 75, 131; Rolls Royce Motors Motor Cars Ltd 34

The BMW 328, which combined excellent performance and durability with superb handling. Owner: Betty Haig.

The following pictures were taken specially for Octopus Books:

Laurie Caddell 124-125; Paul Davies 135; Ian Dawson 1, 4-5, 11, 12, 13 above and below, 14, 15, 21 below, 31, 33, 36, 37, 40, 45, 50, 52, 62, 85, 88, 89, 94, 97, 105, 106, 109, 125, 126, 129, 141, 142, 143, 148-149 bottom, 158; John Lamm 43, 44, 54-55, 137; Chris Linton 127, 148-149 top, 150, 151; Andrew Moreland 121; Octopus Library 32, 41 below, 84, 86 below, 86-87, 101, 138-139; Rainer Schlegelmilch 35, 120; Nicky Wright 2-3, 7, 24, 27, 51 left and right, 56-57, 63, 64-65, 66-67, 67, 68, 69, 70, 71, 73, 74, 77, 78, 79, 136, 140, 160

The author and publishers would also like to thank Sue Forster for her editorial work.